SPATIAL INFORMATION SYSTEMS

General Editors

P. H. T. BECKETT   **M. F. GOODCHILD**
P. A. BURROUGH   **P. SWITZER**

Reactive Data Structures for Geographic Information Systems

PETRUS JOHANNES MARIA VAN OOSTEROM

OXFORD UNIVERSITY PRESS

1993

Oxford University Press, Walton Street, Oxford OX2 6DP
Oxford New York Toronto
Delhi Bombay Calcutta Madras Karachi
Kuala Lumpur Singapore Hong Kong Tokyo
Nairobi Dar es Salaam Cape Town
Melbourne Auckland Madrid
and associated companies in
Berlin Ibadan

Oxford is a trade mark of Oxford University Press

Published in the United States
by Oxford University Press Inc., New York

British Library Cataloguing in Publication Data
Data available

Library of Congress Cataloging in Publication Data
Data available
ISBN 0-19-823320-5

1 3 5 7 9 10 8 6 4 2

Set by Hope Services (Abingdon) Ltd.

Printed in Great Britain
on acid-free paper by
Bookcraft Ltd.
Midsomer Norton, Avon

Spatial Information Systems

Organized societies have collated their knowledge about the earth's surface and its people, animals, and plants for a very long time. They kept their records as texts, tables, diagrams, and maps, which people could read easily but which were difficult to analyse. Now with the aid of computers these data can be stored automatically and retrieved at will to be displayed in a wide variety of ways. Modern spatial information systems also include many logical and mathematical capabilities for abstraction, generalization, re-expression, recombination, rescaling, aggregation and disaggregation, superposition, summerization, interpolation, and error handling that allow thorough analyses that were previously impossible. These analyses not only provide new insights into existing situations but, by using fictional data from simulation models, they enable past and future situations to be modelled and explored. These new tools are revolutionizing mapping, and the inventory and management of our environment. Today automated information systems are used in many branches of science, in business and commerce, in local and national governments, and in international agencies. The applications range from the utilitarian—the mapping of the network of telephones, electricity supply, and sewers—to the esoteric and futuristic, such as in modelling the possible effects of climatic change.

The rapid development in automation has outstripped the supply of scientists and technicians trained to use it. Universities and polytechnics the world over are therefore establishing courses on spatial analysis and geographic information systems, both for their own students and to educate staff already in employment. One of our aims is to provide texts for such courses, for third-year undergraduates, for postgraduates, and for practitioners in the field. The texts are intended to be interdisciplinary, covering basic principles that can be applied in many fields.

New developments in spatial information systems are reported only in the proceedings of conferences and in scientific journals. A second aim is to provide scholarly monographs, written by experts, to describe both the theory and practice in a well-ordered way. Their authors will gather new knowledge from diverse sources and make it widely available and intelligible.

One of the most striking features of a modern information system is that when its full power is used its results rarely fail to surprise and delight. This brings the danger, of course, that the user will be over-impressed by the sheer technology. A further aim, therefore, is to ensure that users are not only impressed but are also delighted because they have achieved a deeper understanding of the world around them.

P. A. Burrough

Utrecht

Preface

Over the past decade, geographic information systems (GISs) have become quite popular and their use is still growing. Examples of applications based on GISs include: topographic and cadastral mapping; hydrographic and aeronautic charting; demography and environmental planning; automated mapping/facility management, command, control, and communication systems; war-gaming and other simulations based on terrain data; ship, aircraft, and car navigation systems.

Advantages of GISs

The ever increasing availability of hardware such as digitizers, scanners, workstations, graphic displays, printers, and plotters for the input, processing, and output of geographic data only partly explains the growing interest in GISs. The major reason for the success of GISs is that they also offer more and new functionality to the map producers and to the map users in comparison with the traditional paper maps. Once a good environment has been created, the maintenance and production of maps becomes easier, because, instead of 'manually redrawing' the complete map, it is sufficient to enter the updates in the geographic database. For the users the advantage of a GIS over the paper map is that they can interact with the system: all kinds of selections based on spatial properties and on thematic attributes can be made, the system can perform various analyses and calculations, and the results are promptly reported in an appropriate graphic or alpha-numeric form. Another advantage of a GIS is that many users can work with the same geographic database simultaneously.

Problems with GISs

As a result of the growing GIS popularity, new applications with more demanding functional requirements are on their way. One of the most fascinating issues is the quest for applications that are based on scaleless, seamless, geographic databases. Furthermore, as data-capturing techniques mature, the available geographic data sets become larger and larger, and are now in the order of gigabytes in the case of vector data and terabytes in the case of raster data. These two developments require GISs with enhanced capabilities. Hardware speed-ups alone are not sufficient, because the requirements are increasing even faster. The purpose of the research described in this book is to create an environment in which new, efficient GISs can be developed efficiently. These two objectives, efficient GISs and efficient development, are reflected in the two major parts of this book.

Part I. Reactive Data Structures

Part I is preceded by Chapter 1, which gives an introduction to GISs: the GIS terminology is defined, some example GISs are presented, and the problem domain is analysed. The chapter concludes with a set of requirements for the data model of future GISs. The strategy followed to reach the goals is a computer science approach. In Parts I and II several new solutions are presented in order to meet these requirements. Prototypes are implemented and used to evaluate the suggested solutions. An important characteristic that distinguishes a GIS from most business data processing systems is the necessity for spatial capabilities. The geographic data model can be regarded as the engine of a GIS. The design of an efficient GIS requires a powerful engine, one that provides sufficient thrust for interactive application. Part I is about such engines: reactive data structures. These are vector-based structures tailored to the efficient storage and retrieval of geometric objects at different levels of detail. Before this goal is reached in Chapter 6, however, Chapter 2 gives an overview of known spatial data structures, Chapters 3 and 4 give some new geometric data structures, and Chapter 5 presents a novel data structure with detail levels. Throughout this book statements can be found like: 'The data structure has good spatial properties.' This is a short formulation of the more elaborate statement: 'When using this data structure, algorithms can solve the spatial queries within the allowed maximum response times.' Similar short formulations are also used in other situations, e.g. in conjunction with the terms 'data model' and 'algorithm'.

Part II. Persistence and Object-Oriented Modelling

Part II is about the development environment for a complete GIS. That should, of course, include the engine described above. In addition, such an environment should also offer sufficient data modelling and database-like (persistence) capabilities. As GISs are more demanding in this area than business data processing systems, relational databases are inadequate. In order to solve the inadequacies, a geographic extension to the relational system is presented in Chapter 7, which includes geometric data types, spatial index structures, and new operators in the query language. The implementation of this extension is described in Chapter 8. However, several drawbacks remain. For example, complex objects, which occur quite frequently in GISs, are hard to represent in the flat tuples of a relational system. Therefore, another approach is taken in Chapter 9, where the object-oriented (OO) paradigm is used as a starting-point. In the OO programming language 'Procol' complex objects and instances of different size, e.g. polygons, pose no problem, as will be shown in a case study. Furthermore, the OO approach has several other well-known advantages, such as the inclusion of abstract data types. The OO environment is well suited to offer libraries of standard objects useful for GIS applications, e.g. user-interface tools, spatial search structures, geometric calculation packages, and so on. The last chapter of this book describes the inclusion of persistent objects in Procol. This precludes the need for storing the data in a file or database system and at the same time solves the problems associated with transferring data to and from the persistent storage system.

Topics Outside the Scope of this Book

It was impossible to cover all aspects related to the current GIS technology. In order to avoid possible confusion, the following list summarizes some topics that remained outside the scope of this book:

- Raster-based GISs, which are especially suited to deal with scanned maps or remote sensing images (Annoni *et al.* 1990, Barker 1988, Burrough 1986, Goodenough 1988, Greenlee 1987, Peuquet 1981*a*, 1981*b*, 1983, USA-CERL 1988).
- Error, quality, and accuracy models for spatial data (Burrough 1986, 1987, Carter 1989, Chrisman 1987, Goodchild 1987*a*, Goodchild and Min-hua 1988, Heuvelink *et al.* 1989, Turner 1988), which are very useful when dealing with data of multiple sources.
- The design of the user-interface of a GIS: the windowing system (Scheifler 1989, Thibault and Naylor 1987), the spatial query language (Egenhofer and Frank 1988, Frank 1982, Goh 1989, Goodchild 1987*b*, Güting 1988*a*, Ingram and Phillips 1987, Joseph and Cardenas 1988), the graphic input and output primitives (Adobe Systems Incorporated 1985, 1987, ANSI 1985, ISO 1986*a*, 1989*a*, Teunissen 1988, Teunissen and van den Bos 1988, 1990, van den Bos *et al.* 1984), and the (mental) psychological system model (Andriole 1986, Reisner 1988).
- The use of rule-based and knowledge-based systems in GIS for: name placement (Doerschler 1987, Freeman and Ahn 1984, 1987, Jones 1990), map generalization (Brassel and Weibel 1988, Müller 1990*a*, 1990*b*, 1990*c*, Nickerson and Freeman 1986, Nickerson 1987, Richardson 1988), or solving other GIS problems (Buisson 1989, Peuquet 1984, Ripple and Ulshoefer 1987, Smith *et al.* 1987*a*, Wu *et al.* 1989).
- Three-dimensional GISs as used in: digital elevation models, 3D cartographic techniques, and geology (Falcidieno and Spagnuolo 1991, Franklin and Lewis 1978, Jones 1989, Kraak 1986, 1988, Lee 1991*b*, 1991*a*, Molenaar 1990). The third dimension may also be used for the time-axis in GISs that deal with historical data (Stonebraker and Rowe 1986, Vrana 1989).
- The development of exchange standards for data types ranging from general to specific: ASCII, relational, graphic, CAD, cartographic, hydrographic, elevation data (Anderson 1989, DMA 1977, 1986*a*, 1986*b*, DCDSTF 1988, DGIWG 1991, Feeley and O'Brien 1989, ISO 1986*b*, Morrison 1989, Nagel *et al.* 1980, US GeoData 1985, Werkgroep SUF-2 1987).

As these topics are closely related to the discussions in this book, it will sometimes be necessary to take a small sidestep and briefly discuss one of them. This is especially true for Chapter 1, which provides a context for the research described in this book. Although the emphasis is on GIS applications, many techniques proposed are also useful in other types of spatial, graphic, pictorial, or geo-information systems, such as: computer graphics, computer-aided design and manufacturing (CAD/CAM), computer vision, and robot systems.

Acknowledgements

The basis of the work described in this book was performed at the Department of Computer Science of Leiden University, and within the framework of the research programme of the Physics and Electronics Laboratory (TNO-FEL). This work was done in the context of a Ph.D. dissertation under the supervision and indispensable guidance of Professor Jan van den Bos, and was concluded in December 1990 (van Oosterom 1990b). More resent research performed at the Physics and Electronics Laboratory is reflected in this book through numerous new references, a few new sections, and one new chapter (Chapter 8). Furthermore, known errors in the Ph.D. thesis were corrected.

The information concerning the typical GIS applications in Section 1.5 was kindly provided by Hido Velsink and Harry Uitermark, both of Dienst van het Kadaster en de Openbare Registers, Ministerie VROM (the LKI system, Subsection 1.5.1 and Fig. 1.7), and by Herman Fernhout of Philips Car Stereo (the CARIN system, Subsection 1.5.3 and Fig. 1.9). The map fragments (Fig. 1.1, Fig. 1.2, and Fig. 6.1) are the copyright of the Dutch Topographic Service. Figs. 1.6 and 3.7 were created by Menno Kraak (Ormeling and Kraak 1987).

Parts of the research described in the book were performed in co-operation with one or more of the following people: Jan van den Bos, Eric Claassen, Chris Laffra, Marcel van Hekken, Tom Vijlbrief, and Marco Woestenburg. The 'presentation of census data' example (see Chapter 9 and Section 10.7) was implemented by Vincent Schenkelaars in Persistent Procol. Many valuable comments and suggestions on the earlier versions of parts of this book were made by the following people: Peter Burrough, Wil Eelsing, Peter Essens, Oliver Günther, Hans Jense, Martien Molenaar, Frans Peters, Donna Peuquet, Andre Smits, Paul Strooper, and Remco Veltkamp.

P. J. M. v. O.

Woudenberg
January 1993

Contents

1 Introduction to Geographic Information Systems

This chapter gives an overview of geographic information systems (GISs) with emphasis on the supporting geographic data model. First, some definitions of a GIS and the data types encountered in GISs are given in Section 1.1. The subsequent section explains why this book concentrates on vector-based GISs instead of raster-based GISs. The general architecture of a GIS is described in Section 1.3. Section 1.4 covers, in more detail, the operations that must be available. Some typical GIS applications are presented in Section 1.5. The motivation for this is twofold: to illustrate the material in this chapter, and because some of the examples will be used in the remaining chapters. Though outside the scope of this book, three-dimensional GISs should not be omitted in an introduction to GIS and are therefore briefly discussed in Section 1.6. The requirements for a general geographic data model are summarized in the last section of this chapter.

Since the early 1980s, there has been a growing interest in the application of GISs. The GIS technology has many applications. A major advantage of such a system is that, although a geographic data set may be presented on a display in much the same way as the traditional paper map (see Figs. 1.1 and 1.2), the user can interact with the system. Section 1.4 gives an overview of the operations that should be available to the user in the interaction process with a GIS. To make it both possible and efficient to answer queries from the user within acceptable response times, a GIS has to be based on an appropriate data model and data structures.

The purpose of the research described in this book is the design of a suitable geographic data model which can be used by an application programmer to develop a specific GIS. In this chapter I describe, illustrate, and justify the requirements for such a data model. The integrated storage and presentation of spatial and thematic data play an important role in this model. To present these data, graphics packages could be used (ANSI 1985, ISO 1989a). However, their data storage facilities are limited: for the most part, only graphic data can be stored; only storage at execution time (or storage in a metafile (ISO 1986b)) is possible; and limited capabilities are available for structuring the data.

Database packages (network, hierarchical, or relational database management systems (DBMSs) (Date 1981, Ullman 1982)) can be used for long-term and structured storage. The drawback of these packages is that they do not handle geographic data very well, because they support only

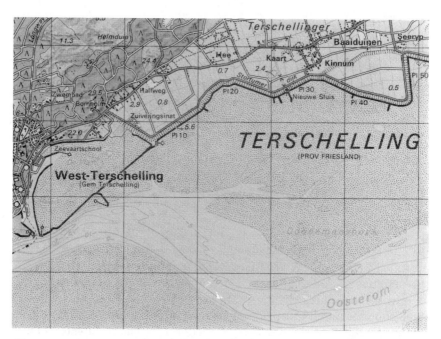

FIG. 1.1. Part of a 1:50,000 Map (©Topographic Service, Emmen)

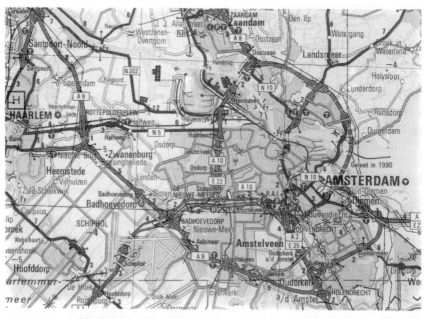

FIG. 1.2. Part of a 1:250,000 Road Map (©Topographic Service, Emmen)

one-dimensional search structures. It is impossible to state queries like: 'Which cities with a population greater than 100,000 lie within 10 kilometres from the river Rhine?' in the data manipulation languages of these DBMSs. Even if it were possible to state these types of queries, they could not be answered efficiently, because the DBMSs lack the proper multidimensional search structures. This explains why many commercial GISs have a dual architecture. The thematic information is stored in a (relational) database and the spatial information is stored in a separate subsystem capable of dealing with spatial data and spatial queries. By means of a unique identification (id), the two components of a geographic entity are linked together. In addition to being inelegant conceptually, this dual architecture also has practical disadvantages: the ids must be kept consistent; 'mixed' multi-dimensional search structures (e.g. two dimensions used for the spatial location and the third dimension used for a thematic attribute) are impossible; and performance is reduced, because objects have to be retrieved and compiled from components that may be stored far apart. Therefore, it is worth while developing a data model that does not have this dual architecture.

1.1 Definitions

Before the definitions of the data types that can be found in a GIS are given, the term GIS itself must be defined. According to the work of the International Geographical Union (IGU), a geographic information system is 'the common ground between information processing and the many fields utilizing spatial analysis techniques' (Tomlinson 1972). This quite general definition of a GIS is refined by Burrough (1986) as 'a powerful set of tools for collecting, storing, retrieving at will, transforming, and displaying spatial data from the real world'. A similar definition is given by the National Center for Geographic Information and Analysis (NCGIA) (1987), which defines a GIS as 'a computerised database management system for capture, storage, retrieval, analysis, and display of spatial (locationally defined) data'. Cowen (1988) puts more emphasis on the integration aspect: 'A GIS is best defined as a decision support system involving the integration of spatially referenced data in a problem solving environment.' Van den Bos (personal communication, 1990) defines a GIS as an instance of a spatial information system. A spatial information system is an information system that integrates and displays thematic and spatial (geometric, topological) data.

From these definitions, it should be clear that the data model plays a central role in a GIS and that the current GISs are the result of the combined efforts in many disciplines, as argued by several authors (Burrough 1986, Cowen 1988, Fisher and Lindenberg 1989, Goodenough 1988, Guptill 1989*a*, Logan and Bryant 1987, Robinove 1986, Smith *et al.* 1987*b*) and visualized in a striking manner on the cover of the *International Journal of Geographical Information Systems* (see Fig. 1.3). The geographic entities

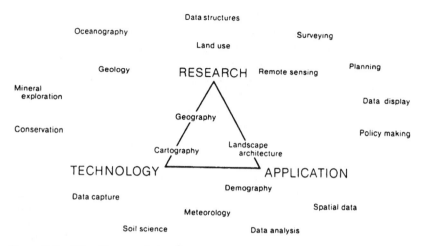

FIG. 1.3. The Cover of the *International Journal of Geographical Information Systems*

or objects in a GIS are based on two different types of data: spatial and thematic. In turn, spatial data have two components: geometric and topological. In this book these terms are defined as follows.

Geometric data have a quantitative nature and are used to represent coordinates, line equations, etc. Basically, there are two formats: raster and vector. The two-dimensional vector format has three subtypes: point (position, 0-cell), polyline (line, arc, chain, 1-cell), and polygon (area, region, 2-cell). Henceforth, these subtypes are called 'geometric primitives'. Of course, other primitives such as circles and splines might be possible, but this is usually not the case in GISs. Though not used in this book, sometimes geometric data are subdivided into positional (locational) data and shape data.

Topological data describe the relationships between the geometric data. There are several types of topological relationship: connectivity, adjacency, and inclusion. Examples of queries concerning topological relationships are: which areas are neighbours of each other (adjacency), which polylines are connected and form a network of roads (connectivity), and which lakes lie in a certain country (inclusion). Topological data are not always stored explicitly, because in principle they can be derived from geometric data. Note that this definition of topological data is slightly different from the stricter one used in mathematics.

Thematic data (application-dependent attributes) are alpha-numeric data related to geographic entities, e.g. the name and capacity of a road. Thematic data may be any kind of data that can be found in the traditional databases, such as strings, integers, and reals.

Fig. 1.4 summarizes and depicts the hierarchy of terms related to geographic data. Note that this figure does not depict a GIS architecture, but

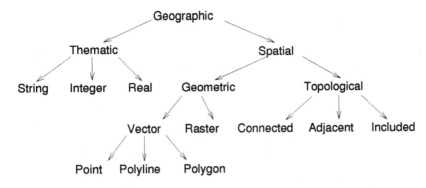

FIG. 1.4. The Hierarchy of Terms related Geographic Data

it does show the terms associated with the different aspects of geographic data. The fact that, for example, the terms 'thematic' and 'spatial' are represented by separate subtrees does not imply that thematic and spatial data should be stored apart. In fact the opposite is true: the data must be stored together in an integrated manner (see Chapter 8).

A geographic database is a set of geographic entities. The geographic database is often organized in layers, which each describe certain aspects of the mapped real world. Another term frequently encountered in the context of GISs is 'graphic', which refers to the visualization of the geographic entities. In Part I, 'Reactive Data Structures', an overview of spatial data structures is given. In addition to well-known data structures for representing geometric and topological data, several new data structures are presented. Thematic data play no role in Part I, but are considered in Part II, 'Persistence and Object-Oriented Modelling'.

1.2 Raster-Based Systems

This section explains why the book concentrates on the vector format instead of the raster format for the geometric data. Note that this discussion is about the internal representation and not about the display system. An advantage of the raster format is that it is appropriate for data-capture techniques like remote sensing (satellites), scanning of existing documents (maps or aerial photographs), and image processing in general, which all deal with raster data. The raster format has some good spatial properties, inherent in the direct addressing of the pixels. Some applications of these properties are as follows:

- A certain (rectangular) region can be selected by using all the pixels with the x and y indices in proper range.
- The neighbours of a pixel can be found by using all the pixels that have indices one lower or higher in comparison with the original pixel.
- If each polygon is represented by a (connected) set of pixels with the same attribute value and every polygon has a different value, then the

point-in-polygon test can be established by retrieving the attribute value of the pixel.

- If two maps are based on the same raster, then map overlay can be performed by a simple pixel-by-pixel algorithm. (This algorithm is also well suited for hardware implementation.)

The fact that a raster algorithm is simple, however, does not mean that it is faster than a more complex vector algorithm, because raster data have the tendency to be very voluminous. Although the raster format possesses good spatial properties, it has some serious disadvantages. Different sources of raster data use different rasters. (Sometimes a pixel does not represent a square but a rectangle.) What is worse, a rectangular raster may not even be rectangular when using a different projection. The geometric data should not depend on a specific projection; that is, a common coordinate system must be used, e.g. the geographic coordinate system.

The main drawback of the raster format is that it basically deals with digital images and not with geographic entities that can be easily displayed and manipulated individually. Therefore, it is more difficult to link the geometric component in a natural way with the other data types of the geographic entities.

In spite of these arguments, a GIS should possess the capability to use raster data as well as vector data, because a user often has no control over the format in which the data are delivered. A GIS should support the conversions between the raster and vector format (Tomlinson 1972). In addition, raster data can be incorporated in a vector-based GIS by treating these data as colour or texture functions on polygons. For example, a satellite image may be projected on a digital elevation model.

1.3   The GIS Architecture

A GIS is more than just a data model in which different types of data structures must be incorporated. In this section important aspects like the database, the user-interface, and exchange standards will be considered. There seems to be a consensus in the literature about the architecture of a GIS. Burrough (1986), Goodchild (1987b), Guptill (1988), Smith *et al.* (1987b), and Tomlinson (1972) describe the following five components: data input, storage, analysis, output, and the user-interface. The way these five components are implemented distinguishes a GIS from other information systems. As pointed out by Tomlinson, the fact that GISs have to deal with spatial or locational data separates them from other types of information systems. Fig. 1.5, taken from (Burrough 1986), illustrates the relationships between the GIS components.

Data input and verification. Some possible sources for the data input are: sensors, existing maps, and field observations. The data can be entered and verified by using one or more of the available tools: scanners, digitizers, graphic displays driven by appropriate software. Format conversion, error detection and editing, topology reconstruction, generalization,

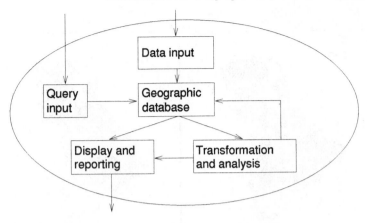

FIG. 1.5. The Five Component Architecture of a GIS

and registration are sometimes also part of the data input process. The cost of data-capture in GISs is usually high. An obvious solution to reduce this cost is multiple use of the same data set. Unfortunately, there are some practical problems: what categories of data should be collected, how accurate should they be, which conceptual structure should be used, and what transfer format should be used? There are several exchange standards available that try to solve some of these problems, including DLMS (DTED, DFAD) (DMA 1977, 1986a, 1986b), USGS DLG (US GeoData 1985), SUF-2 (Werkgroep SUF-2 1987), IGES (Nagel *et al.* 1980), and GKS Metafile (ISO 1986b). One important effort is 'The Proposed Standard for Digital Cartographic Data' (DCDSTF 1988, Morrison 1989) of the Digital Cartographic Data Standards Force (DCDSTF), in which the US Geological Survey plays an important role. In the hydrographic context there are also several ongoing efforts: CEDD (Anderson 1989) of the International Hydrographic Organization (IHO) and MACDIF (Feeley and O'Brien 1989), a Canadian effort. International agencies, such as ISO, NATO, IEEE, and IHO, play an important role in the standardization process.

Data storage and database management. Tasks of this part of the system include traditional DBMS facilities such as support of multiple users and data bases, efficient storage and retrieval, non-redundancy of data, data independence of applications, security, and integrity. To store the thematic data (attributes), a relational DBMS can be used. By using a standard relational DBMS, the portability of the GIS is enhanced. However, the database should also be used for the storage of the geometric and topological aspects of the geographic entities. Standard DBMSs are not suited for this task.

Data transformation, manipulation, and analysis. This part of a GIS is traditionally regarded as the most important one. Of course, analysis is impossible or not useful without the other components of the system. The following analysis operations can be used in many GISs: geometric calcu-

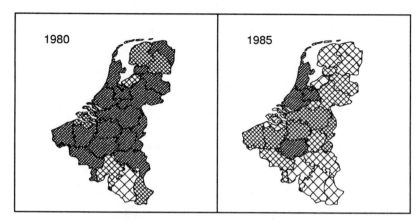

FIG. 1.6. A Multi-Map User-Interface

lations, map overlay computation, network analysis, and several forms of simulation. These operations will be described in more detail in Section 1.4.

Data output and presentation. This can be performed with the help of several different hardware devices, such as displays, plotters, and printers. A number of different presentation formats are available for this purpose: maps, tables, and figures. By using a standard graphic software package, the output part of the GIS becomes less device-dependent and quite a lot of programming may be avoided. There are several packages available: CGI (ISO 1986a), GKS (ANSI 1985), GKS-3D (ISO 1985, Puk and McConnell 1986), Hirasp (Teunissen 1988, Teunissen and van den Bos 1988, 1990), PostScript (Adobe Systems Incorporated 1985, 1987), PHIGS (ISO 1989a) and PHIGS+ (ISO 1989b).

User-interface management. The user-interface is an important layer that is situated between the core of the system, the other four components, and the user. In order to be useful, the user-interface should be uniform and therefore requires a careful design. The operations (Section 1.4) are meant to be applied in the first place by the users. If they can not exploit them, the GIS will not be used to its full extent. Some interesting issues in the user-interface design are: the multi-map user-interface (Fig. 1.6), using a windowing system like X/11 (Scheifler 1989), and the graphic query language, which must be a mixture of a traditional database query language and graphic interaction techniques, complemented with geographic operators. Note that the user-interface consists of an input (queries) and an output (results) part.

1.4 Operations in a GIS

This section describes the operations that should be available to the user of a GIS. Most, but not all, operations described in this section are part of the analysis subsystem of a GIS; see Section 1.3. Two groups of operations are identified. The first consists of the operations that must be present in every GIS; these are the fundamental operations. The second group consists of the operations specific to a certain application, the additional operations. All operations are meant to be used in an interactive mode by the user. This section examines all the fundamental operations and gives a few examples of operations from the second group. The spatial data structures (Part I) should allow for efficient implementation of the operations.

There are three types of fundamental operations: display a map; select entities from a map; and perform spatial calculations with one, two, or more geographic entities as operands. These three operations will be discussed in Subsections 1.4.1–1.4.3. Some additional operations will be treated in Subsections 1.4.4–1.4.6.

1.4.1 Displaying the Map

The map display operation is an obvious GIS operation and seems trivial: visualize what is stored in the database. This is not true, however. First, a number of more complicated steps have to be taken. The type of projection (Evenden 1990, Snyder 1987) has to be chosen. A different example is the required transformation in a navigation system: the display can be north-up or heading-up.

A GIS contains several layers of information, so it has to be decided which layers will be used. Furthermore, the region that is to be displayed and the scale to be used must be selected. After the map has been drawn the user may want to move to an adjacent part of the map, called panning. It is also possible that the user will want to take a closer look at a part of the map, the zoom-in operation. In this case not only are the objects enlarged on the display, but also more details are drawn. The reverse operation of zoom-in is of course zoom-out, where details are removed. This type of zoom-in (and zoom-out) operation will be called logical zoom.

In many situations it is possible to generate different kinds of maps using the same data or derivations thereof. To produce a thematic map, one could use (Ormeling and Kraak 1987): a choropleth, a dot-chart, an iso-line map, a trend surface, a prism map (Franklin and Lewis 1978), etc., to visualize a specific theme. Some GISs require a very powerful display operation; for example IDECAP (van den Bos *et al.* 1984); see Section 1.5.

1.4.2 Selecting Entities

There is a strong relationship between the select and the map display operation. On the one hand, the entities selected are to be displayed on the screen. On the other hand, entities may be selected from those already displayed on the screen.

The selection queries in a GIS include those found in a traditional DBMS. However, the user may also want to pose spatial queries and hybrid queries, the latter being a combination of the traditional and the spatial queries. Traditional queries are stated in an alpha-numeric way using a query language like Structured Query Language (SQL) (Date 1981). In GISs, selection criteria may also be based on topological relationships and on geometric calculations (see Subsection 1.4.3); for instance: 'select all polygons with area greater than 2 square kilometres', or 'select all municipalities adjacent to the city of Leiden'.

There are two basic types of spatial queries: those based on geometric properties of an object and those based on topological properties. Queries dealing with adjacency, connectivity, and inclusion are topological spatial queries; queries dealing with coordinates, area, length, and other sizes are geometric spatial queries. Spatial queries can be made more easily by using the graphic input device pick. The pick device itself has two parameters: shape (the pick primitive, usually a point or a rectangle) and aperture (the maximum distance between the pick primitive and the primitives that are to be selected). Geometric queries that are based on the locations of the geographic entities are called proximity queries. A few examples of proximity queries are: range query, e.g. 'Select everything that overlaps the search area, which can be a specified rectangle, circle, or polygon', and nearest-neighbour query, e.g. 'Find the nearest-neighbour of a given geographic entity'.

1.4.3   Geometric Calculations

This subsection describes geometric calculations of general interest: distance, circumference, area, and centre of gravity. These calculations are used in the select or display operations. Some examples: the area of a polygon is used to convert absolute data (e.g. number of inhabitants of a region) into relative data (population density); the centre of gravity is used for the position of a label in a polygon. The Euclidean distance between point $p_1 = (x_1, y_1)$ and point $p_2 = (x_2, y_2)$ is:

$$D_E(p_1, p_2) = \sqrt{(x_1 - x_2)^2 + (y_1 - y_2)^2}$$

Another useful distance function is the City Block or Manhattan distance:

$$D_C(p_1, p_2) = |x_1 - x_2| + |y_1 - y_2|$$

Formulas for calculating the shortest distance between a point and a polyline or a polygon are a little more complicated. The same applies to the distance between two polylines, two polygons, and also between a polyline and a polygon. The circumference of a polygon defined by points p_1 to p_n (with coordinates (x_i, y_i) for i from 1 to n and $p_{n+1} = p_1$) using Euclidean distance is:

$$C = \sum_{i=1}^{n} D_E(p_i, p_{i+1})$$

The area can be computed by accumulating a running balance of triangular areas, resulting in the formula (when the nodes are numbered in counterclockwise order) (Foley and van Dam 1982):

$$A = \frac{1}{2} \sum_{i=1}^{n} (x_i y_{i+1} - y_i x_{i+1}) = \frac{1}{2} \sum_{i=1}^{n} (x_i - x_{i+1})(y_i + y_{i+1})$$

If it is not known whether the nodes are numbered clockwise or counterclockwise, the true area is the absolute value of A. The centre of gravity (x_c, y_c) is computed similarly:

$$x_c = \frac{1}{6A} \sum_{i=1}^{n} (x_i + x_{i+1})(x_i y_{i+1} - y_i x_{i+1})$$

$$y_c = \frac{1}{6A} \sum_{i=1}^{n} (y_i + y_{i+1})(x_i y_{i+1} - y_i x_{i+1})$$

1.4.4 Map Combination

The ability to combine different maps is one of the main strengths of a GIS. The first variant of this operation is the visual combination of different maps using the map display operation with the correct projection and scale. An example of this is: first draw a polygonal map with soil types, then draw the network map of roads, and finally draw the point map representing the cities. A special type of the visual map combination operation is the display of a map projected on a digital elevation model. A difficulty arises when two polygonal maps, covering the same region, are to be combined, as the second map would obscure the first. The second map could be drawn in a semi-transparent mode, but this is not always a satisfactory solution.

The second variant of map combination operates on two polygonal maps. It first computes the resulting map and then displays the result. This is the polygon overlay problem, and several algorithms are described in the literature (Bentley and Ottmann 1979, Chazelle and Edelsbrunner 1992, Dougenik 1979, Doytsher and Shmutter 1986, Frank 1987, Franklin 1990, Franklin and Wu 1987, Kriegel *et al.* 1991, Mairson and Stolfi 1988, Orenstein 1991, Pullar 1991, van Roessel 1990, Wu and Franklin 1987). Because polygon overlay requires time-consuming computations, it might be useful to store frequently used map overlays (Frank 1987). Note that polygon overlay is also used to solve the problem: 'compute the total area of sandy ground that lies within 2 kilometres of a road'. A buffer zone is created around the road resulting in a polygon map, which is used to perform the overlay with the soil type map.

1.4.5 Network Analysis

For map layers with a network topology, there are several analysis operations. Some examples are as follows (Lahaije 1986, Lupien *et al.* 1987, Moreland and Lupien 1987):

- *Shortest path* Calculate the shortest path between an origin and destination. The minimum path problem is a modification hereof which minimizes the sum of the weighted values of the edges, rather than their geometric length.

- *Location of service centre* Determine the best location of a service centre, e.g. a post office or a bus stop. Try to minimize the total (or maximum) distance between the service centre and the homes of the inhabitants in that region.
- *Travelling salesman* Determine the best or shortest route for a salesman who has to visit a number of cities.

The problems described above and other operation research problems may require a lot of computation, especially if the size of the problem is large, e.g. the number of cities a salesman has to visit. Sometimes, therefore, an almost-best solution requiring less computation time is preferred.

1.4.6   Simulation

The last additional GIS operation addressed here is simulation. One type of the simulation operations resembles the network analysis operations, because they too use the network topology. The network is represented by a weighted graph. The weight of an edge represents a physical property, e.g. the capacity of a road. In simulation, the time aspects and the probability model play important roles. In this respect they differ from network analysis. Examples of simulation are: the evacuation of an area by road, the flow of liquids or gases through pipes, etc. (van de Scheur and Stolk 1986, van Schagen 1987).

Other types of simulation do not use the network topology, but are based on a contiguous space, for example the prediction of soil erosion or noise problems. A raster-based data structure is most appropriate for this kind of simulation. In the soil erosion example, there must be different map layers to represent the sampled soil, land use, rainfall, terrain elevation, and so on.

1.5   Some Typical GIS Applications

Apart from giving an impression of some typical GIS applications, the purpose of this section is to introduce some examples which can be used in the remaining chapters. In each application area, a number of characteristic applications are given, one of which will be described in more detail. The following application areas are recognized: mapping, demography, navigation, (urban) planning, natural resource management, simulation, and analysis. Note that this categorization is quite arbitrary; others are imaginable. Each of the application areas puts a different emphasis on the components of the architecture. Of course, the other GIS components must at least be available in some basic form.

1.5.1   Mapping: The Cadastral Information System

The main purpose of mapping-oriented GISs is to produce geographic databases. Therefore, the data input and storage components are more important than usual. Some examples of mapping-oriented GISs are: topo-

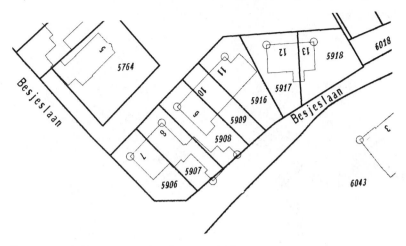

FIG. 1.7. A Fragment of a Cadastral Map

graphic and cadastral mapping, hydrographic and aeronautic charting, land information systems (LIS), and automated mapping/facilities management (AM/FM). In this subsection the 'Cadastral Information System' (in Dutch: Landmeetkundig en Kartografisch Informatiesysteem; LKI) of the Dutch government is described. LKI is the geographic counterpart of the Automatisering Kadastrale Registratie (AKR) system, which contains alpha-numerical legal facts.

Besides parcels (property boundaries), the LKI also contains topographic features. The purpose of LKI (Velsink 1990) is to provide an efficient system for the maintenance of large-scale cadastral and topographic data sets (1:500 and 1:1,000 in urban regions, 1:2,000 elsewhere), see Fig. 1.7. These data sets can be used as a basis for other GIS applications such as: site planning, policy-making, and facilities management. Parcel numbers will be used to link LKI with the system parts of AKR. LKI is based on a network database and a two-dimensional index structure that resembles the Field-tree (Subsection 2.1.7), used for efficient geometric searching. Updating the geographic database is done by first selecting and locking a part of the central database. This part is then downloaded to a workstation, where the updates are processed. Once finished, the updated data set is transferred back to central storage and the lock is released. At the moment, about 10 per cent of the cadastral maps and about 20 per cent of the topographic maps of the Netherlands are stored in LKI. The complete digital map is expected to be finished by the year 2005. The expected size of the total database is about 40 gigabytes.

1.5.2 Demography: Presentation of Census Data

The development of information systems to analyse and present census data has a long history. The emphasis in this kind of GIS is on the output and user-interface parts. To process the 1990 US census data, the US Bureau

[a] Municipalities   **[b] Provinces**

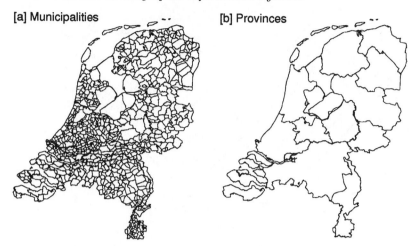

FIG. 1.8. Municipalities and Provinces in the Netherlands

of the Census has developed the sophisticated TIGER system (Boudriault 1987, Kinnear 1987, Marx 1986), the successor of the famous topologically structured DIME (US Bureau of the Census 1970) system. The case presented in this subsection is inspired by the first interactive GIS developed in the Netherlands. This system is called IDECAP (Projectgroep 1982, van den Bos *et al.* 1984) and its development started in the early 1970s at the University of Nijmegen by Van den Bos *et al.* IDECAP presents raster-based thematic data on a vector-based topographic base map. The user first graphically selects a region of interest and then selects one or two census variables. The result is promptly presented in the form of a convenient map.

In the case of presentation of census data, the entities are typically the administrative units for which the census data are collected, unlike IDE-CAP, where the census data are collected in a raster format. Fig. 1.8 [a] shows the smallest available administrative units in the Netherlands, called municipalities (Dutch: *gemeenten*). The geometric attribute of the municipality is a polygon, corresponding to the boundary line of the municipality.

In the Netherlands there are five levels of administrative units for which census data are collected. These are, listing from small to large: municipality, economic geographic region (Dutch: *economisch geografisch gebied*), nodal region (Dutch: *nodaal gebied*), corop region (Dutch, no translation available), and province (Dutch: *provincie*). These units are hierarchically organized. The provinces, the largest administrative units, are shown in Fig. 1.8 [b]. The different administrative units are natural candidates for the division into detail levels.

Each administrative unit interacts with the others and, of course, contains the geometrical shape and the relevant census data, e.g. number of inhabitants. There are two types of relationship. The first type describes

FIG. 1.9. The CAR Information and Navigation System

the topological adjacency, the second describes the hierarchic aggregation structure. Until now, we have discussed only the data aspect of the entities. The operations aspect of the system must provide procedures for: displaying according to the selected census data, reporting census data to the user and other parts of the system in forms and tables (because of the inherent hierarchy, the top-level units may want to inquire their lower-level units and aggregate the information), and entering new census data.

1.5.3 Navigation Systems: The CARIN System

The GISs in this category are used to control ships, aircrafts, cars, or other vehicles. Although the systen is built on top of a geographic database, it is not necessary for the user ever to see a map generated by the system. The user may receive audio commands, or the information system might even control the vehicle in an autonomous mode. These GISs are often connected with sensors, for example to receive signals from satellites of the Global Positioning System (GPS) (Bakker *et al.* 1986, Stansell 1983) in order to determine the position of the vehicle. In this subsection, a car navigation system will be presented.

In 1984, Philips Research started the development of an electronic co-pilot that is able to guide the motorist to his or her destination, determine the current position, and warn the driver about fog-banks, traffic jams, and accidents. The co-pilot was called CARIN (Lahaije 1986, Thoone 1987), an abbreviation for CAR Information and Navigation system. Fig. 1.9 shows the CARIN system in practice. CARIN consists of the following hardware components: car CD-I player, position location system, board computer, sensors, car radio, screen, keyboard, and voice synthesiser. The

CARIN system has the following tasks: route planning, route guiding, route following, visualization, and information retrieval. The position location system is based on dead reckoning. This system compares the input from two wheel (left and right) sensors and an electronic compass with the digital road map. A problem arises if the car is transported by train; in this case the driver needs to enter the new position. In the future, CARIN could be linked into the GPS system, which would also solve the train-position problem.

The CARIN system does not show a map when the car is moving, because this would distract the driver too much. The system only shows direction changes on a small display in a simple format. It is backed up by a voice synthesizer able to repeat the instruction for the driver who may not have time to take in visual instructions. As the CARIN system uses a CD-I player for the storage of geographic data, it is important to minimize the number of disk reads, because a CD-I player has a relatively long seek time. This can be done by using an appropriate storage structure.

1.5.4   Planning, Simulation, and Analysis: The Geographic Analysis Intelligence System

This subsection covers a large category of applications. The reason for this is that planning often requires simulation and analysis in order to evaluate possible decisions. Some examples of GISs in this category are: decision support in emergency planning (van de Scheur and Stolk 1986), war-games (van Schagen 1987), and management of utilities such as electricity, water, gas, and telephone. The emphasis, in these types of planning systems, is on the analysis component. The case presented here is a Command, Control, and Communication Information (C^3I) system. It is called the Geographic Analysis Intelligence System, and is similar to a C^3I system that is currently under development for the intelligence organizations at the Royal Netherlands Air Force (RNLAF) airbases. Two of the main tasks of the intelligence organization on an airbase are:

- to supply various kinds of information to pilots for mission preparation;
- to supply the battlestaff of an airbase with scenario information, e.g. current and previous activities of own and enemy troops.

It is evident that the information provided (target data, enemy weapon system data, airspace management data, battle flow data, etc.) must be accurate and up to date. Maintenance of these data becomes more and more time-consuming, especially in times of tension and war, when the data tend to change very frequently. To support the tasks of the airbase intelligence organization, the Geographic Analysis Intelligence System must perform the following functions:

- Reliable and fast storage of intelligence data (target information, enemy weapon systems information, planning lines) and airspace management data (flying routes and operations zones). These data may be entered or updated in various ways, e.g. by the user with a keyboard

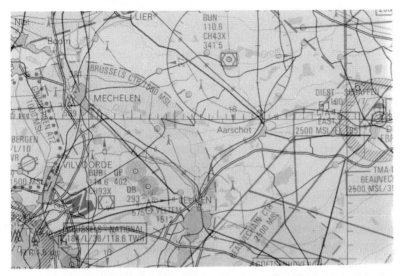

FIG. 1.10. Background Map with Geographic Intelligence Objects

and graphic input devices, by external systems through a network, or by magnetic tape readers.

- Retrieval, presentation, and distribution of previously mentioned intelligence and airspace management information.

Both data entry/updating and data presentation can be alpha-numeric and geographic. In the alpha-numeric case, the data are displayed in tabular form on a screen. Input of new or modified data is done by using an input device such as a keyboard. In the case of graphic data manipulation and presentation, the data are displayed on a background map, using both alpha-numeric characters and graphic symbols. Data are then entered or updated using an input device like a mouse or a light pen in combination with the background map. Both ways of interaction with the data are needed to support the intelligence organization in its main tasks.

For the purpose of this book, we are interested mostly in the geographic requirements of the system, so we will examine them in more detail. The background maps that the system uses must contain several map layers with information about administrative boundaries (borderlines of countries), lines of communication (rivers, roads, railroads), land classification and coverage (swamps, forests), terrain elevation, and names of objects.

Maps of various scales must be available, with the possibility of displaying more information on a more detailed background map. It will be clear that the system must contain large quantities (several gigabytes) of these background map data. The data on the background maps are static; i.e., they change only once every few months. Data that change very often are the geographic intelligence objects that the system must contain. The amount of these data is somewhat smaller than the amount of background data. The intelligence objects can be divided into four data types: point (most targets and threats), circle (airspace management zones and some targets), polyline (planning lines and flying routes), and polygon (airspace management zones and some targets). An example of the presentation of geographic objects on a background map is given in Fig. 1.10. Note that the information density in this example is very high. It is better to present only the relevant map layers selected in interaction with the user.

Another geographic requirement of the Geographic Analysis Intelligence System is the need to store the various levels of detail that can be distinguished in the system. Because background maps of varying scales are used, it is also required that the geographic objects that are displayed on these maps can vary in detail. For instance, when a very schematic background map is used, it is desirable to represent an airbase only as a pin-point location, using a simple symbol, whereas when one uses a very detailed background map, it may be desirable to display more details of the airbase, e.g. runways and shelters. For the purpose of mission preparation, it is important to select only the relevant objects for display, because a very large number of geographic objects are available in the system and displaying them all would confuse the pilot. Relevant objects for the pilot would be the target of the mission, the objects that are near enough to the mission flying route to cause some threat, and relevant airspace management data.

To display and manipulate geographic data, it is very important that these data can be retrieved from and stored into the system fast enough for interactive use. A typical retrieval operation is the selection of geographic objects of one of the types mentioned above within a certain area of interest.

1.6   Three-Dimensional Systems

The meaning of the term 'three-dimensional' may not be as obvious as it may seem in this context. Two well-known interpretations exist. The first is in the sense of a digital elevation model (DEM); that is, for every location (x, y) there is a value z representing the height of the surface. A DEM is sometimes referred to as a 2.5-dimensional model. Second, there is the true three-dimensional interpretation as required for the representation of volumes. A few examples of the latter are: sub-surface modelling, three-dimensional air space management, and three-dimensional cadastral boundaries in tall buildings.

Suitable data models for the true three-dimensional applications are found in CAD or geometric modelling textbooks, e.g. (Mäntylä 1988, Mortenson 1985). There are three basic data models for storing the DEM data, which is sometimes also called digital terrain model (DTM) (Petrie and Kennie 1990):

1. contours: in analogy with height contours on paper maps (Watson 1992);
2. regular grid points: as can be found in DLMS-DTED (DMA 1986*b*);

3. random points: used as input for creating a triangular irregular network (TIN).

Each of these models may be converted into one of the others. These conversions are based on mathematical techniques such as interpolation. Further, variants on these basic models do exist; for example, there is a TIN-based model that also accepts line segments for input and guarantees that the model will represent the elevation at these lines exactly. This is achieved by including the input lines in the set of edges of the triangles.

There are several possibilities for the visualization of DEM data. The most spectacular ones are the perspective side views with shading and hidden surfaces removed. Draping raster or vector data over the DEM may even intensify the effect. Orthogonal projections from above are useful when height visualization techniques are used: slope hatching or shading, colour coding of the height, and height contours with labels. The visualization of true three-dimensional volumes requires other techniques such as removing 'outer volume layers', semi-transparent rendering, or slicing of the volumes (Jense 1991).

In addition to visualization, there are several other essential DEM operations, such as finding local and global extremes, performing slope analysis, cut-and-fill (volume) calculations, and line-of-sight calculations. Another interesting application is that of watershed run-off modelling, which can be used as a basis for soil erosion predictions.

The majority of the techniques presented in this book can be adapted for three- and higher-dimensional use, though they are usually presented in two-dimensional situations. Some examples of three-dimensional applications within a GIS, besides DEMs, are:

- storage of attributes (thematic data) directly with the two-dimensional geometric data, resulting in three or more dimensions;
- time aspects, to add an extra dimension; for example regarding the change in land use.

1.7 Requirements of a Data Model for GISs

The requirements presented in this section summarize the previous sections of this chapter, especially the sections covering the GIS operations and the example GIS applications. As mentioned before, the backbone of a GIS will be the data model based on a spatial data structure. In this section only the requirements that are more or less specific for a GIS will be discussed. The more standard requirements such as 'it must be possible to retrieve objects efficiently based on some of the thematic attributes' will not be elaborated upon; this is comparable with the use of indices in a relational database. The requirements for the GIS data model can be summarized as follows:

r1 Geometric, topological, and thematic data must be stored in a single undivided storage system in order to avoid the drawbacks of a dual architecture; see Section 1.1. Because of their different nature, these three types each require a specific data structure. These data structures have to be tightly integrated in the GIS data model.

r2 The data model must possess good spatial capabilities, allowing efficient implementation of the three fundamental operations described in Section 1.4. Geometric properties are necessary for efficient implementation of operations such as selection of all objects within a rectangle, picking an object from the display, map overlay computations, and so on (Chrisman 1990*b*, van Oosterom 1988*a*, 1988*b*). Topological properties are required for efficiently solving network analysis problems and for topological selections, such as 'select the countries that are adjacent to Germany'.

r3 In the data model it must be possible to store the data in a form suitable for use at several levels of detail. This requirement was also stated by Lohrenz (1988). Recently, there appears to be a growing interest in multi-scale geographic databases (Becker *et al.* 1991, Konečný 1991, NCGIA 1989). There are several reasons for this. First, if too much information is presented to the user at any one time, it will reduce his or her efficiency in perceiving the relevant information, as noted in the old saying, 'You can't see the wood for the trees.' Second, unnecessary detail will increase the access time to the database and slow down the drawing on the display. The latter is especially true for primitives which have become smaller than a pixel after the transformation to screen coordinate system, because these are not visible and thus time is wasted. The user looking at a small-scale (coarse) map must not be bothered with too much detail. When he or she zooms in, additional details must be added. This operation, logical zooming, in contrast to normal zooming which only enlarges, must have at least two additional effects. First, the objects that were already visible in the smaller-scale map now have to be redrawn in finer detail. Second, objects which were not visible in the smaller-scale map may now become visible. Logical zooming is closely related to map generalization (Ormeling and Kraak 1987, Robinson *et al.* 1984, Shea and McMaster 1989), meaning that aggregation/disaggregation must be supported too. The rule of thumb on constant pictorial information density is that the total amount of information displayed on one screen should be about the same for all scales.

r4 In the data model it must be possible to store the data in layers of geographic entities (data planes or overlays), so that it is possible to add or remove a layer from the display. This requirement is quite obvious, because it is natural to organize the way we perceive the world in manageable and comprehensible parts. So, this must also be the case in an information system that models a part of the real world.

r5 The data model must support persistent storage. There should not be a large gap between the data saved in a database or file system and the data stored in data structures of a running GIS program. Persistent storage provides two major advantages. First, it is easier to implement a GIS, because the programmer does not have to worry about saving the volatile data of a running program and vice versa. When analysing existing GIS programs, it becomes clear that significant parts of the code are devoted to this task, especially for parsing the data from the storage system in order to initialize the data structures properly. Second, when good persistent storage is provided the program will become faster, because a significant amount of overhead is avoided. Note that the persistent storage system is still responsible for the traditional DBMS tasks such as support of multiple users, security, and integrity.

r6 The data model must be dynamic, so that it is possible to add, remove, or change geographic entities and entity-types. Not all these operations are equally important. Many GISs are not very dynamic with regard to the geographic data; that is, the number of update transactions (in a fixed time period) is relatively small compared with the size of the whole data base. Once in a while the geographic data are updated by the data supplier. The creation of the initial data structure is not time-critical, but it should result in an efficient data structure. The addition of an object must be performed rapidly and should not decrease the performance of the search operation based on the data structure significantly. Changing and deleting objects will not happen often, so these operations may be less efficient. An object may be deleted by simply setting a flag indicating that this object is no longer present. This also allows an efficient undo.

Part I

Reactive Data Structures

A reactive data structure is defined as a geometric data structure with detail levels, intended to support the sessions of a user working in an interactive mode. The term 'reactive data structure' was introduced by van den Bos in the IDECAP project (1982, 1984). It enables the information system to react promptly to the user's actions, such as the request to present data in a certain spatial region with the appropriate amount of detail. It is difficult to find a data structure that satisfies all the requirements discussed in Section 1.7. It turns out that the combination of requirements $r2$ (spatial capabilities) and $r3$ (levels of detail) forms a major bottleneck. Therefore, one or more requirements are temporarily relaxed. We will concentrate on the spatial capabilities ($r2$) first. As mentioned in Section 1.1, the vector format is used to store the geometric data, that is: points, polylines, and polygons.

The deficiencies of using map sheets in geographic information systems are well known and have been described by several authors (Chrisman 1990a, Frank 1988b). The obvious answer to these deficiencies is a seamless or sheetless database. A seamless database is made possible in an interactive environment by using the proper geometric data structure. A few of the most well-known geometric data structures are described in Section 2.1. A binary tree or B-tree (Bayer and McCreight 1973, Comer 1979, Knuth 1973) is inadequate to represent the spatial aspects because geometric data are multi-dimensional. The same applies to other storage structures based on one-dimensional order, e.g. splay trees (Tarjan 1987).

Additional structures are needed to store the relationships between these objects: topological data (see Section 2.2).

The second requirement, and one that will get considerable attention in this part of the book, is the incorporation of levels of detail (*r3*). Section 2.3 reviews several known data structures with levels of detail. A data structure with the two properties, spatial organization and detail levels, is the basis for a seamless, scaleless geographic database (Guptill 1989*b*). The detail levels must be integrated in the spatial organization. A solution has to be found somewhere in between:

- Try to define a discrete number of levels of detail and store them separately, each with its own geometric data structure. This was done by Waugh and Dowers (1988). Though fast enough for interactive applications, this solution is not particularly elegant. It introduces redundancy because some objects have to be stored at several levels. Apart from the increased memory usage, another drawback is that the data must be kept explicitly consistent. If an object is edited at one level, its 'counterpart' at the other levels must be updated as well.

- Store everything in one geometric data structure, at the most detailed level. Compute generalization when required during the GIS operations. There are several different types of generalization (see Section 2.3). Note that the traditional generalization process can be complicated and involves human decisions. If at all possible, generalization will require too much time to be performed completely in interactive applications.

A reactive data structure is a novel type of data structure catering for multiple detail levels with rapid responses to geometric queries. Reactive data structures may be either static or dynamic. In a dynamic reactive data structure entries can be inserted, deleted, and updated efficiently while the structure remains balanced. In Chapter 3 a novel geometric data structure, in the context of GISs, is suggested: the BSP-tree. The first reactive data structure is also described in that chapter: the BSP-tree with detail levels. It is a static structure. However, requirement *r6* (dynamics) gets considerable attention in the subsequent chapters of Part I.

Chapter 4 describes the KD2B-tree and the Sphere-tree, two new geometric structures. The main characteristic of these geometric data structures is that they are orientation-insensitive. Unfortunately, they do not incorporate detail levels. The subsequent chapter presents the BLG-tree, a novel structure that is used to represent several detail levels of a polyline or polygon. It is not a geometric data structure. As described in Chapter 6, however, the combination of the BLG-tree with a geometric data structure (for example the KD2B-tree), together with other techniques, results in a truly reactive data structure. This approach resulted in the first dynamic and reactive data structure: the Reactive-tree.

2 Overview of Spatial Data Structures

This chapter gives an overview of several important spatial data structures that have been published. As such, it may be regarded as a review paper presenting known material. In addition however, a categorization for spatial data structures is designed. Three categories of data structures are recognized: geometric data structure, topological data structure, and data structure with detail levels. These are described in Sections 2.1–2.3, respectively. Only data structures requiring linear storage space in practice are considered, and even the associated constant must be quite small, i.e. preferably not larger than 2. Higher storage space requirements are not realistic, because of the large amounts of data in GISs. So the Range-tree (Bentley and Friedman 1979, Overmars 1983), for example, a geometric data structure for storing n points that requires $O(n \log n)$ storage space, is not included, in spite of its good query time: $O(\log^2 n + t)$ for a range query where t points are found. In this chapter only the two-dimensional variants of the data structures are described. Many of these structures can also be used in three-dimensional applications.

2.1 Geometric Data Structures

In the last section of Knuth's famous book on sorting and searching (1973), the topic of multi-attribute retrieval is introduced. On page 554 he gives an example of an orthogonal range query, or range query for short, to select 'all cities with ($21.49° \leq$ latitude $\leq 37.41°$) and ($57.72° \leq$ longitude $\leq 70.34°$)'. In spite of the fact that Knuth gives a few data structures for multi-attribute data, he states: 'No really nice data structures seem to be available for such orthogonal range queries.' That was back in 1973. By now, several solutions for storing and retrieving multi-dimensional geometric data efficiently have been proposed in literature. It is impossible to cover all published geometric data structures. This section presents a selection of structures to illustrate the different techniques used. The following data structures will be briefly described: KD-tree, Quadtree, Conjugation-tree, R-tree, Two-dimensional ordering, Grid File, Field-tree, and Cell tree. Some other interesting structures, which are not described in this section, are: BANG file (Freeston 1987, 1989), Buddy Hash-tree (Kriegel *et al.* 1989), Buddy-tree (Seeger and Kriegel 1990), EXCELL (Tamminen 1983), hB-tree (Lomet and Salzberg 1989), LSD tree (Henrich *et al.* 1989), R-file (Hutflesz *et al.* 1990), and Twin Grid Files (Hutflesz *et al.* 1988). The reason that these are not included in this book is not that they are less efficient, but that only the most fundamental structures could be described and these latter structures are often derived from the structures described.

The research presented in this book resulted in new geometric data structures: the multi-object BSP-tree and the generalized KD-tree described in Chapter 3, and the KD2B-tree and the Sphere-tree described in Chapter 4. General evaluation criteria that should be kept in mind, while studying the geometric data structures, are:

- Which of the point, polyline, and polygon entities can be stored and retrieved efficiently?
- Which search operations can be performed—exact point match, range queries, polygon queries, nearest-neighbour, etc.?
- Is the data structure based on rectangular or non-rectangular division of space?
- Are the geometric primitives themselves used for divisions or are arbitrary regular divisions used?
- Is three- or higher-dimensional use possible?
- Are non-zero sized entities, such as polygons, sometimes partitioned or not? In the latter case we call the structure geometrically object-oriented.
- Is it possible to insert, delete, and update entries while the structure remains balanced? That is, is it a static or dynamic structure?
- Is the data structure suited for secondary memory? That is, are the characteristics of secondary memory, such as disk-paging, taken into account?
- Are the geometric primitives stored inside the data structure or are they indexed (stored elsewhere and referred to by a pointer)?
- What is the worst-case performance for building and querying the structure?
- How complicated or simple are the data structure and its algorithms (e.g. recursion) to implement?
- What is the performance in GIS practice with large irregular data sets?

Several other related evaluation rules are described in literature, for example by Frank and Barrera (1989), and by Kriegel *et al.* (1989).

2.1.1 The KD-tree

The overview of geometric data structures starts with the KD-tree. Bentley (1975) described this binary tree structure. The basic form of the KD-tree stores K-dimensional points. This section concentrates on the two-dimensional (2D) case. Each internal node of the KD-tree contains one point and also corresponds to a rectangular region. The root of the tree corresponds to the whole region of interest. The rectangular region is divided into two parts by the x-coordinate of the stored point on the odd levels and by the y-coordinate on the even levels in the tree (see Fig. 2.1). A new point is inserted by descending the tree until a leaf node is reached. At each internal node the value of the proper coordinate of the stored point

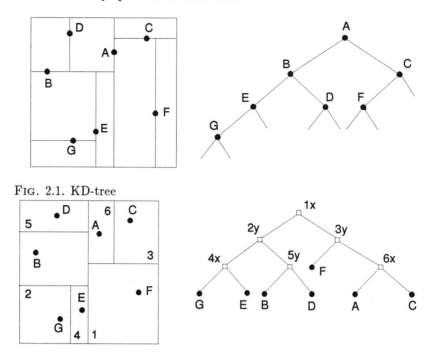

FIG. 2.1. KD-tree

FIG. 2.2. Adaptive KD-tree

is compared with the corresponding coordinate of the new point and the proper path is chosen. This continues until a leaf node is reached. This leaf also represents a rectangular region, which in turn will be divided into two parts by the new point. The insertion of a new point results in one new internal node.

A disadvantage of the KD-tree is that the shape of the tree depends on the order in which the points are inserted. In the worst-case, a KD-tree of n points has n levels. The adaptive KD-tree (Bentley and Friedman 1979) solves this problem by choosing a splitting point (which is not an element of the input set of data points), which divides the set of points into two sets of (nearly) equal size. This process is repeated until each set contains one point at the most (see Fig. 2.2). The adaptive KD-tree is not dynamic: it is hard to insert or delete points while keeping the tree balanced. The adaptive KD-tree for n points can be built in $O(n \log n)$ time and takes $O(n)$ space for $K = 2$. A range query takes $O(\sqrt{n} + t)$ time in 2D where t is the number of points found. In the remainder of this book, the term 'adaptive KD-tree' is usually abbreviated to KD-tree. Another variant of the KD-tree is the bintree (Tamminen 1984). Here the space is divided into two equal-sized rectangles instead of two rectangles with equal numbers of points. This is repeated until each leaf contains one point at the most.

The modification that makes the KD-tree suitable for secondary memory is described by Robinson (1981) and is called the KDB-tree. For practical

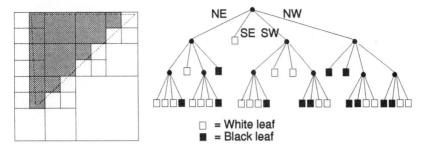

FIG. 2.3. Region Quadtree

use, it is more convenient to use leaf nodes containing more than one data point. The maximum number of points that a leaf may contain is called the bucket size. The bucket size is chosen in such a way that it fits within one disk page. Also, internal nodes are grouped and each group is stored on one page in order to minimize the number of disk accesses. Robinson describes algorithms for deletions and insertions under which the KDB-tree remains balanced. Unfortunately, no reasonable upper bound for memory usage can be guaranteed.

Matsuyama *et al.* (1984) show how the geometric primitives polyline and polygon may be incorporated using the centroids of a bounding box in the 2D-tree. Rosenberg (1985) uses a 4D-tree to store a bounding box by putting the minimum and maximum points together in one 4D point. This technique can be used to generalize other geometric data structures that are originally suited only for storing and retrieving points. The technique works well for exact match-queries, but is often more complicated in the case of range queries. In general, geometrically close 2D rectangles do not map into geometrically close 4D points (Hutflesz *et al.* 1990). The ranges are transformed into complex search regions in the higher-dimensional space, which in turn result in slow query responses.

2.1.2   The Quadtree

The Quadtree is a generic name for all kinds of trees that are built by recursive division of space into four quadrants. Samet (1984, 1989) gives an excellent overview, of which this subsection is a partial abstract. The best known Quadtree is the region Quadtree, which is used to store a rasterized approximation of a polygon. First, the area of interest is enclosed by a square. A square is repeatedly divided into four squares of equal size until it is completely inside (a black leaf) or outside (a white leaf) the polygon or until the maximum depth of the tree is reached (dominant colour is assigned to the leaf) (see Fig. 2.3). The main drawback is that it does not contain an exact representation of the polygon. The same applies if the region Quadtree is used to store points and polylines. This kind of Quadtree is useful for storing binary raster data, but, as mentioned in Section 1.1, in this book I concentrate on data structures based on the vector format.

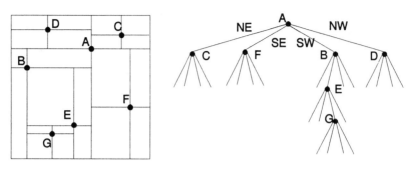

FIG. 2.4. Point Quadtree

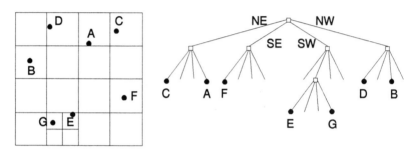

FIG. 2.5. PR Quadtree

The Quadtrees in the remainder of this subsection store vector format representations. The point Quadtree resembles the KD-tree described in the previous subsection. The difference is that the space is divided into four rectangles instead of two (see Fig. 2.4). The PR Quadtree (point region) does not use the points of the data set to divide the space. Each time, it divides the space, a square, into four equal subsquares, until each contains one point at the most (see Fig. 2.5).

The extensive research efforts on Quadtrees in the last decade resulted in many more variants. For example, there is the PM Quadtree for storing a polygonal map, and the CIF Quadtree, which is a Quadtree particularly suited for rectangles. Descriptions of them can be found in Rosenberg (1985) and Samet (1984). The KD-tree and the various Quadtrees are well-known structures and their efficiency has been compared for several different input data sets (Beckley *et al.* 1985, Matsuyama *et al.* 1984, Rosenberg 1985).

2.1.3   The Conjugation-tree

The Conjugation (or Ham-sandwich) -tree is a search structure for 2D points and it was designed to solve the query: 'How many points are contained in a specified half-space?' It was introduced by Edelsbrunner and Welzl (1986). Edelsbrunner (1987) also describes how the Conjugation-tree can be used to determine which points lie on a line, or in a polygon. The Conjugation-tree is based on the geometric notions 'bisect' and 'conjugate'.

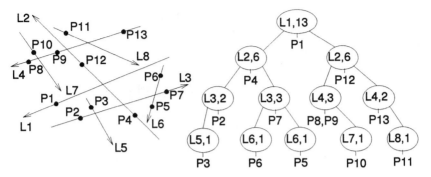

FIG. 2.6. A Conjugation-tree for Thirteen Points

A directed line l bisects the set S, which contains n points, if both open half-spaces defined by l contain no more than $\lfloor n/2 \rfloor$ points. This means that line l contains at least one point if n is odd. The set that contains the points on l is called $S_{on}(l)$ and the sets that contains the points to the left and to the right of l are called $S_{left}(l)$ and $S_{right}(l)$, respectively. Line l' is a conjugate of l for S if l' simultaneously bisects $S_{left}(l)$ and $S_{right}(l)$. The fact that l' exists and that it can be found in $O(n)$ time is proved in Edelsbrunner (1987).

Unless S is empty, the Conjugation-tree has a root that contains the bisector l, the size of the set $|S|$, an array that stores $S_{on}(l)$ in sorted order, and two pointers (left and right) to the subtrees. Both these subtrees have bisector l', but are conjugation subtrees for different sets; $S_{left}(l)$ and $S_{right}(l)$ (see Fig. 2.6, taken from Edelsbrunner and Welzl 1986). The Conjugation-tree takes $O(n)$ space and results from the definition of bisector that the height of the tree is bounded by $\lceil 2\log n \rceil$. As the conjugate l' can be found in $O(n)$ time, the tree can be built in $T(n) = O(n) + 2T(n/2) = O(n\log n)$. The time required to determine the number of points contained in the query halfplane is:

$$Q(n) = Q(n/2) + Q(n/4) + O(\log n) = O(n^{2\log\frac{1+\sqrt{5}}{2}}) = O(n^{0.695})$$

The binary search in the sorted array takes $O(\log n)$ time. Additionally, at most three out of four grandchildren of a common node can be intersected by the border of the halfplane. The Conjugation-tree is also used for reporting points in the halfplane. If t points are found, this takes $O(n^{0.695} + t)$ time. Other query regions, such as polylines or polygons, are also possible by a simple change of the search algorithm, while maintaining the same bounds for the query time $Q(n)$.

In summing up, its advantage over the PR Quadtree is that the Conjugation-tree is balanced, resulting in better worst-case query times. Its advantage over the KD-tree is that in the Conjugation-tree a query line (possibly the border of a halfspace or polygon) intersects at most three out of four grandchildren of a common node, which results in better worst-case query

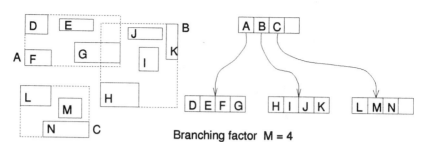

FIG. 2.7. R-tree

times. The major drawbacks of the Conjugation-tree are that it is a static structure and that only points can be stored.

2.1.4 The R-tree

The R-tree is an index structure that was defined by Guttman in 1984. The leaf nodes of this multi-way tree contain entries of the form: (I, object-id), where object-id is a pointer to a data object and I is a bounding box (or an axes-parallel minimal bounding rectangle, MBR). The data object can be of any type: point, polyline, or polygon. The internal nodes contain entries of the form: (I, child-pointer), where child-pointer is a pointer to a child and I is the MBR of that child. The maximum number of entries in each node is called the branching factor M and is chosen to suit paging and disk I/O buffering. The Insert and Delete algorithms of Guttman assure that the number of entries in each node remains between m and M, where $m \leq \lceil M/2 \rceil$ is the minimum number of entries per node. An advantage of the R-tree is that pointers to complete objects (e.g. polygons) are stored; so the objects are never fragmented.

Fig. 2.7 shows an R-tree with two levels and $M = 4$. The lowest level contains three leaf nodes and the highest level contains one node with pointers and MBRs of the leaf nodes. Coverage is defined as the total area of all the MBRs of all leaf R-tree nodes, and overlap is the total area contained within two or more leaf MBRs shortciteD25B. In Fig. 2.7 the coverage is $A \cup B \cup C$ and the overlap is $A \cap B$. It is clear that efficient searching demands both low coverage and overlap: a packed R-tree.

Roussopoulos and Leifker (1985) describe the Pack algorithm which creates an initial R-tree that is more efficient than the R-tree created by the Insert algorithm. The Pack algorithm frequently uses a nearest-neighbour function during the creation of the packed R-tree. The city block distance function might be used because it is computationally cheap and favours pairs of points or MBRs that lie along the main axes instead of diagonal orientations, resulting in rectangles with less coverage.

The R$^+$-tree (Faloutsos *et al.* 1987), a modification of the R-tree, avoids overlap at the expense of more nodes and multiple references to some objects (see Fig. 2.8). Analytical results indicate that R$^+$-trees allow more efficient searching, in the case of larger objects. The efficiency of the R-tree

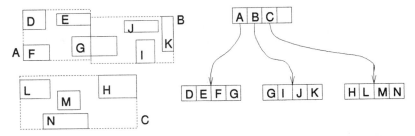

FIG. 2.8. R^{+}-tree

and the R^{+}-tree has been compared with several other secondary memory data structures (Beckmann *et al.* 1990, Faloutsos *et al.* 1987, Frank and Barrera 1989, Greene 1989, Günther and Bilmes 1991, Kriegel *et al.* 1989, Roussopoulos and Leifker 1985, van Oosterom and Claassen 1990).

2.1.5   Two-Dimensional Orderings

Quite a different geometric data structure is based on the ordering of points in a discrete two-dimensional space. This technique is also called tile indexing, and it transforms a two-dimensional problem into a one-dimensional one. Then a well-known data structure for one-dimensional storage and retrieval is used, e.g. the B-tree (Bayer and McCreight 1973). Besides storing discrete points, the ordering technique can also be used for cells of a gridbased representation, for example in combination with a linear Quadtree (Abel and Smith 1983, Samet 1984), Grid File (Nievergelt *et al.* 1984), or Field-tree (Frank and Barrera 1989). The presentation in this subsection is based on several papers (Abel and Mark 1990, Goodchild and Grandfield 1983, Jagadish 1990) and on the book of Samet (1989), in which the following seven orderings are described: row, row prime, Morton, Hilbert, Gray code, Cantor-diagonal, and spiral.

The row ordering simply numbers the cells row by row, and within each row the points are numbered from left to right (see Fig. 2.9 [a]). The row prime (or snake-like, or boustrophedon) ordering is a variant in which alternate rows are traversed in opposite directions (see Fig. 2.9 [b]. Obvious variations are column and column prime orderings in which the roles of row and column are transposed.

Bitwise interleaving of the two coordinates results in a one-dimensional key, called the Morton key (Orenstein and Manola 1988). The Morton key is also known as Peano key, or N-order, or Z-order. For example, row $2 = 10_{bin}$ column $3 = 11_{bin}$ has Morton key $13 = 1101_{bin}$ (see Fig. 2.9 [c]). Hilbert ordering is based on the classic Hilbert–Peano curve, as drawn in Fig. 2.9 [d]. Gray ordering is obtained by bitwise interleaving the Gray-codes of the x and y coordinates. As Gray codes have the property that successive codes differ in exactly one bit position, a 4-neighbour cell only differs in one bit (see Fig. 2.9 [e]) (Faloutsos 1988). In Fig. 2.9 [f] the Cantor-diagonal ordering is shown. Note that the numbering of the points

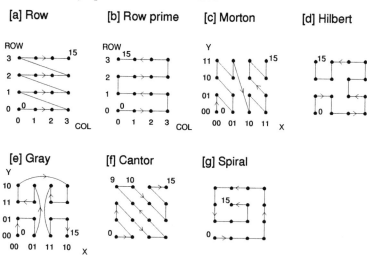

FIG. 2.9. Seven Different Orderings

is adapted to the fact that we are dealing with a space that is bounded in all directions; e.g., point (1,3) got order number 10 instead of 11 and point (3,3) got 15 instead of 24. Finally, the spiral ordering is depicted in Fig. 2.9 [g]. Mark (1990) has identified three desirable properties of spatial orderings:

1. An ordering is continuous if, and only if, the cells in every pair with consecutive keys are 4-neighbours.
2. An ordering is quadrant-recursive if the cells in any valid Quadtree subquadrant of the matrix are assigned a set of consecutive integers as keys.
3. An ordering is monotonic if, and only if, for every fixed x, the keys vary monotonically with y in some particular way, and vice versa.

Ordering techniques are very efficient for exact match queries for points, but less efficient for other types of geometric queries, e.g. a range query. Abel and Mark (1990) conclude from their practical comparative analysis of five orderings (they do not consider the Cantor-diagonal and the Spiral orderings) that the Morton ordering and the Hilbert ordering are, in general, the best alternatives. Goodchild (1989) proved that the expected difference of 4-neighbour keys of an n by n matrix is $(n+1)/2$, which is exactly the same as in the case of row and row-prime ordering.

2.1.6 The Grid File

The principle of a Grid File is the division of the space into rectangles (regular tiles, grids, squares, cells) that can be identified by two indices, one for the x-direction and one for the y-direction. The Grid File is a non-hierarchical structure. The geometric primitives are stored in the grids, which are not necessarily of equal size. There are several variants of this

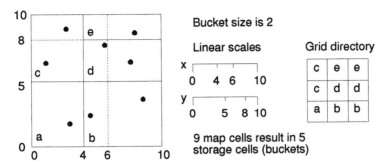

FIG. 2.10. The Grid File

technique. In this subsection the file structure as defined by Nievergelt *et al.* (1984) is described.

The advantage of the Grid File as defined by Nievergelt is that it adjusts itself to the density of the data unlike most other Grid Files, but it is also more complicated. The cell division lines need not be equidistant; for x and y there is a one-dimensional array (in main memory) with the linear scales, the actual sizes of the cells. Neighbouring cells may be joined into one bucket if the resulting area is a rectangle. The buckets have a fixed size and are stored on a disk page. The grid directory is a two-dimensional array, with a pointer for each cell to the correct bucket. Fig. 2.10 shows a Grid File with linear scales and grid directory. The Grid File has good dynamic properties. If a bucket is too full to store a new primitive and it is used for more than one cell, then the bucket may be divided into two buckets. This is a minor operation. If the bucket is used for only one cell, then a division line is added to one of the linear scales. This is a little more complex but still a minor operation. In case of a deletion of primitives, the merging process is performed analogous to the splitting process for insertion.

2.1.7 The Field-tree

The Field-tree resembles both the PR Quadtree and the Grid File but, more important, it has a number of other characteristics. In addition to points, it is suited to store polylines and polygons in a non-fragmented manner. The Field-tree is the only geometric data structure described in this section that takes account of the fact that there may exist geometric objects of varying importance (see Section 2.3). But it is unclear whether this can be realised in a dynamic environment. During the last decade several variants of the Field-tree have been published by Frank *et al.* (1983, 1989, 1986). In this subsection attention will be focused on the Partition Field-tree. It is interesting to note that the Field-tree is used by the land registry office in the Netherlands; see Subsection 1.5.1.

Conceptually, the Field-tree consists of several levels of grids, each with a different resolution and a different displacement (see Fig. 2.11). A grid

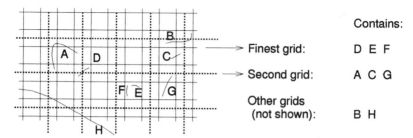

FIG. 2.11. The Positioning of Geometric Objects in the Field-tree

cell is called a field. Actually, the Field-tree is not a hierarchical tree, but a directed acyclic graph, as each field can have one, two, or, four ancestors. At one level the fields form a partition and therefore never overlap. In another variant, the Cover Field-tree (1989), the fields may overlap. It is not necessary that at each level the entire grid explicitly be present as fields.

A newly inserted object is stored in the smallest field in which it completely fits (unless its importance requires it to be stored at a higher level). As a result of the different displacements and grid resolutions, an object never has to be stored more than two levels above the field size that corresponds to the object size. Note that this is not the case in a Quadtree-like structure, because there the edges at different levels are collinear. The insertion of a new object may cause a field to become too full. In that case an attempt is made to create one or more new descendants and to reorganize the field by moving objects down. This is not always possible. A drawback of the Field-tree is that an overflow page is sometimes required, as it is not possible to move relatively large or important objects of an overfull field to a lower level field.

2.1.8   The Cell tree

Günther designed the Cell tree (Günther 1988, Günther and Bilmes 1988, 1991) to facilitate searches on polygonal objects. The Cell tree can be viewed as a combination of the BSP-tree (see Chapter 3) and the R^+-tree (Faloutsos *et al.* 1987). Each Cell tree node corresponds to a convex polygon. Each internal node is partitioned into at least m cells (except for the root) in a manner analogous to the BSP-tree; see Fig. 2.12, taken from Günther and Bilmes (1991). The leaf nodes, which are all on the same level, contain all information that may be required to answer a query. This is an advantage over the R-tree and other structures using MBRs, where data objects have to be retrieved from secondary memory. Each node of the Cell tree corresponds to one disk page.

Data objects, which may be arbitrary polygons, are represented as unions of convex cells. To insert a new object, each such component is inserted into the Cell tree. The insertion of one convex component may cause the

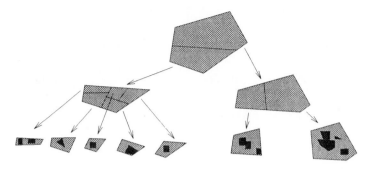

FIG. 2.12. A Cell Tree with Cells Shown in Dark Grey

creation of several new leaf entries, owing to splits. This is a general property of index structures based on partitioning of space into non-overlapping
components, and is also true for the R^+-tree. If a leaf node becomes too
full it is split, and the split propagates upwards in a manner similar to the
R-tree. In the case of many overlapping data objects, the splitting does
not succeed and overflow pages have to be used. If during the deletion of a
data object a leaf node becomes empty, it is eliminated and the elimination
is propagated upwards. Internal nodes with less than m entries are deleted
and the leaf entries under these nodes are reinserted. This may involve
more work than in the corresponding case of the R-tree, because in the
R-tree a whole subtree can be inserted at once.

2.2 Topological Data Structures

The previous section considered a number of geometric data structures.
They are not used to store topological data explicitly. Several special data
structures are known for this purpose. Which topological data structure
should be applied depends on how the geometric primitives are used. If the
geometric primitives are used to represent polygonal areas, then the TIGER
structure is well suited; see Subsection 2.2.1. Subsections 2.2.2–2.2.4 deal
with the topological data structures in the cases of network, contour, and
point data.

2.2.1 The TIGER Structure

The Topologically Integrated Geographic Encoding and Referencing
(TIGER) system was developed by the US Bureau of the Census (Boudriault 1987, Kinnear 1987, Marx 1986). The description of the TIGER
structure is taken from Boudriault (1987). The TIGER structure is the
refined successor of older chain-node structures, such as DIME (US Bureau of the Census 1970), also developed by the US Bureau of the Census,
and POLYVRT (Peucker and Chrisman 1975), developed at the Harvard
Laboratory for Computer Graphics and Spatial Analysis.

When storing polygonal areas, the structure captures the relationships
between the polygon, polyline, and point primitives, which are called 2-

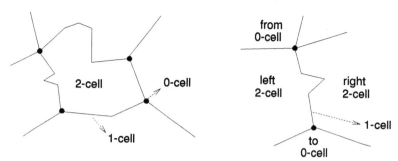

FIG. 2.13. TIGER Structure

cells, 1-cells, and 0-cells respectively. A 1-cell is a polyline, the first and the last points of which are by definition 0-cells. These 0-cells are called from and to 0-cells and they are the same in case of a loop 1-cell. The points of a 1-cell between the 0-cells are called curvature points and are connected by vectors. A 1-cell has two sides, a left and right 2-cell (see Fig. 2.13). The complex of 1-cells, that is the collection of all 1-cells, covers a finite region of the plane. The unbounded region that surrounds the 1-cell complex is represented by a 2-cell labelled with a special code: outside. In addition to these definitions, the topological structure must obey two rules. The rule of Topological Completeness requires that the topological relationships between cells are complete; for example, each 2-cell, except outside, is completely surrounded by a set of 'connected' 1-cells. The rule of Topological–Geometric Consistency requires a consistent relationship between the geometric placement of cells and the pure topological relationships of cells; for example, no two 2-cell interiors share a common coordinate.

2.2.2   The Single-Valued Vector Map

Molenaar (1989) has developed a formal data structure (FDS) for the so-called single-valued vector maps. A similar system, called MINI-TOPO (Wagner 1989), has been developed at the Defense Mapping Agency (DMA, USA). Recently a 3D variant of the FDS has been published (Molenaar 1990), but it will not be described here because it is outside the scope of this book. The FDS incorporates a topological structure that is similar to the TIGER structure. The term 'single-valued vector map' refers to a terrain description based on a set of well-defined conventions, which will be described below. This results in a query space that can be thoroughly analysed. The FDS of a single-valued vector map deals not only with topology, but also with attribute data organized in classes. The primitives node and arc play an important role in the FDS. Fig. 2.14 depicts the FDS for a single-valued vector map. Note that each ellipse represents a set and that each arrow represents a one-to-many relationship between associated sets: $A \rightarrow B$ means that each element of A belongs to at most one element of B. The FDS has to obey the following conventions (Molenaar 1989):

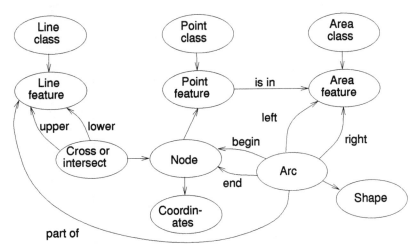

FIG. 2.14. The FDS for a Single-Valued Vector Map

1. The feature classes must be mutually exclusive; this means that each (point, line, or area) feature must have only one class label.
2. A feature class must contain features of only one type.
3. When the map is analysed as a graph, all points used to describe the geometry of a vector map should be treated as nodes.
4. The arcs of this graph must be geometrically represented by segments of straight lines.
5. For each pair of nodes there must be at most one arc connecting them. In addition to that, the nodes may be connected by one or more chains.
6. In the geometric interpretation two arcs may not intersect.
7. For each geometric element there must be at most one occurrence of each of its link types to a feature.

One of the most elegant aspects of the FDS is that it is possible to describe the topological relationships between the features in a formal manner. In this way the semantics of topological terms such as neighbour, island, branching, crossing, intersecting, ending, inside, etc., are defined. More details can be found in Molenaar (1989).

Points, lines, and areas in the FDS correspond to 0-, 1-, and 2-cells in the TIGER structure. The FDS and the TIGER structure can both be used for network data, that is non-polygonal data with only points and lines. The topological capabilities of the FDS are more powerful than those of the TIGER structure. In the TIGER structure, it is not possible to represent a fly-over crossing. This situation occurs in the case of a network of roads or pipelines. The topological structure of this kind of datum should be represented by a graph, as incorporated in the FDS of Molenaar. The points and lines are the nodes and edges of the graph. In the 3D variant of the FDS this problem would not occur, because there is no intersection in the 3D space.

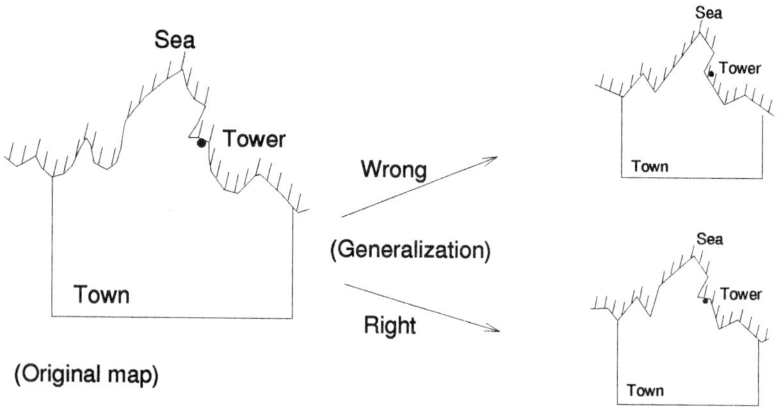

FIG. 2.15. Generalization Error Introduced by the Omission of Topology

Another example that illustrates the topological capabilities of the FDS is shown in Fig. 2.15, where a tower is located within a town adjacent to the coastline. When producing a small-scale representation of this map by using a line generalization algorithm (Section 2.3), it is possible that the tower will end up in the sea. The only way to ensure that the tower remains on land is to store that topological fact, as can be done in the FDS.

2.2.3   The Contour Structure

Another type of data frequently encountered in GISs is based on contours, for example height contours and density iso-lines of a certain physical property. The nesting of the contours is described by a multi-way tree (see Fig. 2.16). A disadvantage of this tree is that there is no upper bound on the number of children per node, which makes it somewhat harder to implement efficiently.

The topological structure of contour data can be used to implement an algorithm for colouring the areas between the contours on a raster display: during the inorder traversal of the tree, completely fill the area inside each contour that is encountered. It is interesting to note that the Digital Feature Analysis Data (DFAD of DLMS (DMA 1977)), which among other things represents polygonal areas, uses as its topological base not a TIGER-like structure, but the contour structure. In case of DFAD, it is not the tree that is stored, but the result of the inorder traversal of the tree.

2.2.4   The Point Structure

Point data sets seem to have no inherent topological structure. They can be based on a regular grid. Basically, point data are used for two purposes in GISs. The first is to represent the arbitrary location of objects, for example a tower. The second is to represent the locations in some regular grid at which certain properties are measured, for example the height of the terrain, the number of people in a grid cell, air pollution. In the case of a regular square grid, the point data degenerate into raster data. As already

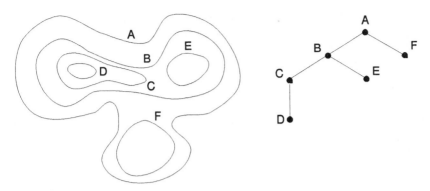

Fɪɢ. 2.16. Contour Structure

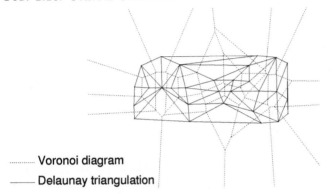

········ Voronoi diagram

—— Delaunay triangulation

Fɪɢ. 2.17. Voronoi Diagram

explained in Section 1.1, rasters have good spatial properties and clear (implicit) topological structures. If the points are stored in a 2D array, the neighbours of a point are found simply by incrementing or decrementing one of the array indices. A useful topological structure for other types of point data is the Voronoi diagram (Preparata and Shamos 1985) (see Fig. 2.17). For the points p_i for $1 \leq i \leq n$, the Voronoi diagram consists of n regions $V(i)$ with the property that, if $(x, y) \in V(i)$, then p_i is the nearest-neighbour of (x, y). If $H(p_i, p_j)$ is the half-plane with the set of points closer to p_i than to p_j, then

$$V(i) = \bigcap_{i \neq j} H(p_i, p_j)$$

The Voronoi diagram is used to answer spatial queries. For example, if we assume that the points represent the locations of postoffices, then the Voronoi diagram can be used to answer the query: 'Which is the closest postoffice when I am situated in location x?' The straight-line dual of the Voronoi diagram is called the Delaunay triangulation (Veltkamp 1988, 1989) and is often used in a conversion from point data to contour data.

This triangulation forms a good basis for interpolation techniques. A defining property of the Delaunay triangulation is that the circumcircle of each triangle does not contain any other data point in its interior.

2.3   Structures with Detail Levels

In this section several data structures with detail levels are described. First, some remarks about the fundamental problems concerning detail levels are made. The concept of multiple detail levels cannot be defined as sharply as that of spatial searching. It is related to one of the main topics in cartographic research: map generalization, that is, the derivation of small-scale maps (large regions) from large-scale maps (small regions). A number of generalization techniques for geographic entities have been developed and described in the literature (Ormeling and Kraak 1987, Robinson *et al.* 1984, Shea and McMaster 1989):

- simplification (e.g. line generalization);
- combination (aggregate geometrically or thematically);
- symbolization (e.g. from polygon to polyline or point);
- selection (eliminate, delete);
- exaggeration (enlarge);
- displacement (move).

Unlike spatial searching in the previous section, which is a pure geometric/ topological problem, map generalization is application-dependent. The generalization techniques are categorized into two groups (Müller 1990*b*, Ormeling and Kraak 1987): geometric and conceptual generalization. In geometric generalization the basic graphic representation type remains the same, but is, for example, enlarged. This is not the case in conceptual generalization, in which the representation changes; e.g., change a river from a polygon into a polyline type of representation.

Generalization is a complex process of which some parts, e.g. line generalization (McMaster 1987, Müller 1987), are well suited to be performed by a computer and others are more difficult. Nickerson (1987) shows that very good results can be achieved with a rule-based expert system for generalization of maps that consist of linear features. Shea and McMaster (1989) give guidelines for when and how to generalize. Müller (1990*c*) also applies a rule-based system for selection (or its counterpart: elimination) of geographic entities. Brassel and Weibel (1988) present a framework for automated map generalization.

Mark (1989), Müller (1990*a*, 1990*b*), and Richardson (1988) all state that the nature of the phenomenon must be taken into account during the generalization—this in addition to the more traditional guidelines, such as the graphic representation (e.g. number of points used to draw a line) and the map density. This means that it is possible for a different technique to be required for a line representing a road than for a line representing a river.

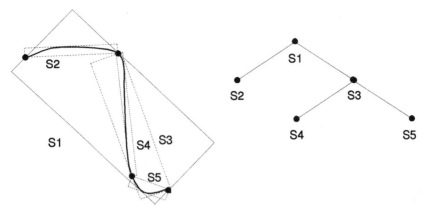

FIG. 2.18. The Strip tree

It is important to note that the data structures with detail levels presented in this book are used to store the results of generalization processes.
In this section four known structures are described: Strip tree, Arc tree,
Multi-Scale Line Tree, and Flintstones. These structures are all related
to line generalization or approximation techniques. The same is true for
our BLG-tree; this novel structure is presented in Chapter 5. The support
of other map generalization techniques is considered in Chapter 6. Chapter 9 briefly discusses how other generalization techniques may fit into an
object-oriented data model.

2.3.1   The Strip tree

In 1981, one of the first data structures with detail levels was described by
Ballard; the Strip tree. The Strip tree is a hierarchical representation for
curves and consists of a binary tree where each node contains a rectangle,
called a 'strip'. Lower levels in the tree correspond to finer resolutions.
The strip for the part of the curve from $p_s = (x_s, y_s)$ to $p_e = (x_e, y_e)$ is
defined by the sextuple $(x_s, y_s, x_e, y_e, w_r, w_l)$, where w_r and w_l, the right
and left distances to the directed line segment $[p_s, p_e]$, are chosen in such
a manner that the strip contains just the curve.

The curve from p_s to p_e is split into two parts: $[p_s, p_m]$ and $[p_m, p_e]$, where
p_m has distance $d = \max(w_r, w_l)$ to the line segment $[p_s, p_e]$. New strips
are defined for these line segments and the splitting process is repeated
until the required resolution w is met: $w_r + w_l < w$. The splitting process
and the strips are represented in the Strip tree (see Fig. 2.18).

Ballard gives basic operations on Strip trees performed at resolution w;
e.g., calculate the length of a curve; display the curve; and intersect two
curves. All these operations are quite straightforward when using the Strip
tree. In the case of length approximation, it must be noted that the actual
length of the curve is larger than the approximation at resolution w and
that it is impossible to give an upper bound for the error. Problems are

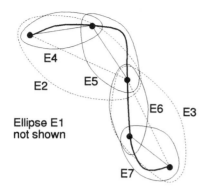

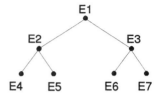

Arc-tree with position
of ellipses indicated.
Note that these are not
actually stored. (Only
points are stored.)

FIG. 2.19. The Arc tree

caused by 'fractal-like' or 'zigzag' shaped curves (see Fig. 5.3 [a] on p. 86). A region is represented by a closed curve stored in the Strip tree. For these trees additional operations are defined: determine whether a point is inside a region; calculate the area of a region; intersect two regions; etc.

2.3.2   The Arc tree

Similar to the Strip tree, the Arc tree also is a binary tree that represents a curve in a hierarchical manner with increasing accuracy in the lower levels of the tree. The Arc tree is described by Günther in his Ph.D. thesis (1988). The major differences with the Strip tree are that no explicit information regarding the bounding primitive (the strip in case of a Strip tree) is stored and the way in which the split point p_m on the curve is chosen. By choosing p_m exactly halfway between the two end-points p_s and p_e, measured along the curve itself, interesting properties can be obtained. Each of the curves has exactly length $l/2$ where l is the length of the original curve. The division of curves is repeated until the required resolution is met (see Fig. 2.19). In the case of the Arc tree this means that the length of a curve has to be smaller than a certain value. The only thing stored in a node of the Arc tree is p_m. Length, start-point, and end-point of the complete curve are stored once elsewhere. Clearly, the Arc tree requires less storage space than the Strip tree.

The kth approximation consists of 2^k line segments each representing a part of the curve with length $l/2^k$. A nice property of the Arc tree is that each part of the curve is contained in the ellipse whose focal points are the start and end-point of the approximating line segment and whose major axis is $l/2^k$. In a manner similar to the Strip tree, closed curves are used to represent regions. Also, hierarchical point inclusion tests and set operations are defined. An advantage of the Arc tree is that an upper bound for the approximation error for the length of the curve can be given. Another advantage is that the Arc tree is always balanced. A special variant, the Polygon Arc tree, is described by Günther (1988) for storing a hierarchical representation of a polygon.

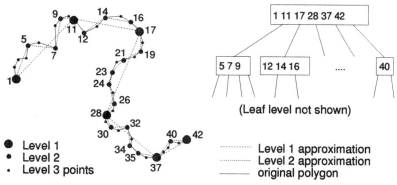

FIG. 2.20. The Multi-Scale Line Tree

2.3.3 The Multi-Scale Line Tree

The Multi-Scale Line Tree is more rooted in cartography, in contrast to the Strip tree and the Arc tree, which both come from computer science. The Multi-Scale Line Tree is a multiway tree, introduced by Jones and Abraham (1986, 1987), which is suited to represent a polyline for a fixed number of scales. It is based on the Douglas–Peucker (1973) line generalization algorithm. The Douglas–Peucker algorithm selects the same points as the Strip tree would for a polyline at the required resolution. However, no strips are stored and no tree is created.

The required scales must be determined in advance. For each required scale or resolution there is one level in the tree. The top level of the Multi-Scale Line Tree is created by applying the Douglas–Peucker algorithm at the most coarse resolution. The next level is created by storing the new points, which are selected by applying the algorithm to the next resolution, in a number of new child nodes. This is done in such a manner that all new points that lie in between two points of the first level, with regard to their original sequence number, are stored in one node. Also, a pointer is made from the root to the new node. This process is repeated for all required scales (see Fig. 2.20). The Multi-Scale Line Tree has the disadvantage that it introduces a discrete number of detail levels, which must be determined in advance. The number of children of a node in the Multi-Scale Line Tree is not fixed. This implies that the tree is not binary, a preferred feature for implementations.

2.3.4 The Flintstones

The Flintstones, recently described by Veltkamp (1990), form a hierarchical representation and approximation scheme for polygons. There is a 2D and a 3D variant of the Flintstones; in this section only the former is discussed. The zero-order approximation of the polygon is the minimal bounding circle, which passes through at least two points of the original polygons (see Fig. 2.21 [a], taken from Veltkamp 1990). The polygon is split into two polylines defined by the touching points. Each polyline is

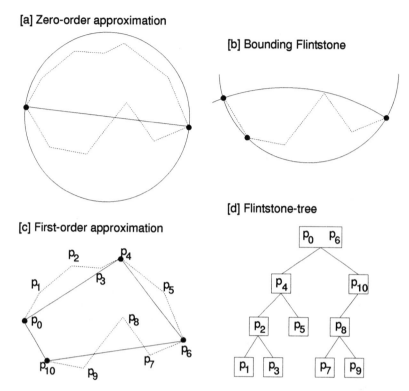

[a] Zero-order approximation

[b] Bounding Flintstone

[d] Flintstone-tree

[c] First-order approximation

FIG. 2.21. The Flintstones

enclosed by a Flintstone, that is, the smallest intersection of two circles. Each polyline is split into two polylines by the point that touches the arc with the largest curvature (see Fig. 2.21 [b]). The start, end, and touch-points form the first-order approximation (Fig. 2.21 [c]). The whole process of enclosing polylines by Flintstones is repeated on the second level and this produces the second-order approximation. This is repeated until the original line segments of the polygon are reached. The whole process is naturally represented in a binary tree (Fig. 2.21 [d]). In a manner similar to that for the Strip tree and the Arc tree, hierarchical operations, such as point inclusion and intersection, can be performed.

Advantages of the Flintstones over the Arc tree are that better split points are chosen (the Arc tree just selects the point halfway, which does not necessarily result in good approximations) and that the spheric tests with Flintstones are computationally cheaper than tests with ellipses. A drawback of the Flintstones is that the tree is not necessarily balanced. It is hard to compare the quality of the points selected by the Flintstones with the points selected by the Strip tree. Practical performance tests will have to tell which structure provides the best approximation.

3 The Reactive Binary Space Partitioning Tree

None of the data structures presented in the overview of the previous chapter combined the geometric capabilities with multiple levels of detail. That is, no reactive data structure was presented. In this chapter our first reactive data structure is introduced. It is a main memory structure and it is a modification of the Binary Space Partitioning (BSP-) tree to include detail levels. The BSP-tree is one of the few spatial data structures that does not organize space in a rectangular manner. A prototype system has been implemented. An important result of this implementation is that it shows that binary space partitioning trees of real maps have $O(n)$ storage space complexity in contrast to the theoretical worst-case $O(n^2)$, with n the number of line segments in the map.

A short description of the original BSP-tree is given in Section 3.1, together with some minor modifications for the GIS environment. The subsequent section shows how the basic spatial operations can be implemented efficiently by using a BSP-tree. Section 3.3 describes the most important difference from the original BSP-tree, the incorporation of detail levels. An application based on the reactive BSP-tree is presented in Section 3.4. The balancing of the BSP-tree is discussed in Section 3.5 for both the static and the dynamic case. Section 3.6 contains the first practical results from our implementation. Finally, the pros and cons are discussed in Section 3.7.

3.1 The BSP-tree and Some Variations

In this section the original BSP-tree is first described. Then two modifications for GIS applications are given: the object BSP-tree and the multi-object BSP-tree. These are storage structures for polygons and polylines. Variations on the BSP-tree for storing points could be developed easily, but are not described here.

3.1.1 The Original BSP-tree

The original use of the Binary Space Partitioning tree was in three-dimensional (3D) computer graphics (Fuchs *et al.* 1983, 1980, Teunissen and van Oosterom 1988). The BSP-tree was used by Fuchs to produce a hidden surface image of a static three-dimensional scene. After a pre-processing phase it is possible to produce an image from any view-angle in $O(n)$ time, with n the number of polygons in the BSP-tree.

In this chapter the two-dimensional BSP-tree is used for the structured storage of geometric data. It is a data structure that is not based on a

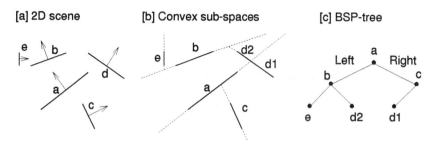

FIG. 3.1. The Building of a BSP-tree

rectangular division of space. It uses the line segments of the polylines and the edges of the polygons to divide the space in a recursive manner. The BSP-tree reflects this recursive division of space. Each time a (sub) space is divided into two subspaces by a so-called splitting primitive, a corresponding node is added to the tree. The BSP-tree represents an organization of space by a set of convex subspaces in a binary tree. This tree is useful during spatial search and other spatial operations. Fig. 3.1 [a] shows a 2D scene with some directed line segments. A 2D scene is used here, because it is easier to draw than a 3D scene. However, the principle remains the same. The 'left' side of the line segment is marked with an arrow. From this scene, line segment a is selected and the 2D space is split into two parts by the supporting line of a, indicated by a dashed line in Fig. 3.1 [b]. This process is repeated for each of the two subspaces with the other line segments. The splitting of space continues until there are no line segments left. Note that sometimes the splitting of a space implies that a line segment (which has not been used for splitting itself) is split into two parts. Line d, for example, is split into d1 and d2. Fig. 3.1 [b] shows the resulting organization of the space, as a set of (possibly open) convex subspaces. The corresponding BSP-tree is drawn in Fig. 3.1 [c]. In the 3D case, supporting planes of flat polygons are used to split the space instead of lines.

The choice of which line segment to use for dividing the space very much influences the building of the tree. It is preferable to have a balanced BSP-tree with as few nodes as possible. This is a very difficult requirement to fulfil, because balancing the tree requires that line segments from the middle of the data set be used to split the space. These line segments will probably split other line segments. Each split of a line segment introduces an extra node in the BSP-tree. Section 3.5 contains a further discussion on the balancing of the BSP-tree.

Fig. 3.2 contains a Pascal-like code of a program that builds a BSP-tree. The program **BuildTree** is a variation of the traditional method (non-incremental) for building a BSP-tree (Fuchs *et al.* 1980). The procedure **SplitLine** and the functions **LinePosition**, **CreateNode**, and **GetLine** are not included, because their implementation is trivial. A node in the

```
Program BuildTree;
type BSP = ^node;
     node = record
                 segm: Line;
                 l, r: BSP
     end;
var  root: BSP;
     newsegm: Line;

function AddLine(tree:BSP; segm:Line): BSP;
var Lsegm, Rsegm: Line;
begin
   if (tree = nil) then tree := CreateNode(tree, segm)
   else case LinePosition(tree,segm) of
        LEFT:  tree^.l := AddLine(tree^.l, segm);
        RIGHT: tree^.r := AddLine(tree^.r, segm);
        SPLIT: SplitLine(tree, segm, Lsegm, Rsegm);
               tree^.l := AddLine(tree^.l, Lsegm);
               tree^.r := AddLine(tree^.r, Rsegm);
   AddLine:=tree;
end;

begin
   root := nil;
   while GetLine(newsegm) do root := AddLine(root, newsegm);
end.
```

FIG. 3.2. Incremental BSP-tree Building Algorithm

BSP-tree is represented by the record type **node**, which contains a line segment and pointers to the left and right child. Initially, the tree is empty. As long as GetLine can fetch a new line segment, it is added to the BSP-tree with a call to the function AddLine. AddLine checks whether the correct position in the BSP-tree is found. This is true if the pointer **tree** in the BSP-tree is **nil**. In that case a new node is created and added to the tree. Otherwise, LinePosition determines in which subtree the line segment has to be stored. The storage of the line segment is implemented by a recursive call to AddLine. It is possible that the line segment has to be split first.

The splitting of line segments has a serious drawback. If we have n line segments in a scene, then it is possible that we end up with $O(n^2)$ nodes (Fuchs *et al.* 1980) in the tree. It will be clear that this is unacceptable in GIS applications, in which we typically deal with 10,000 or more line segments. However, this is a worst-case situation and the actual number of nodes will not be that large; see Section 3.6 for a more detailed discussion.

The BSP-tree is intended for interactive applications in which fast responses are required. Minimizing the memory usage is considered less important and so I have not paid a lot of attention to it. However, the typical memory usage is about three to four times the size of the original map data. There are three reasons for this: the memory space required for pointers, the splitting of line segments, and the replication of points, as the end of one line segment is often the start of another.

3.1.2   The Object BSP-tree

The BSP-tree, as discussed so far, is suited only for storing a collection of (unrelated) line segments. In a modelling system it must be possible to represent a closed object, for example (the interior of) a polygon in the 2D case. The object BSP-tree is our extension to the BSP-tree to cater for object representation. It stores the line segments that together make up the boundary of the polygon. The object BSP-tree has explicit leaf nodes which do not contain line segments to split the subspaces any further. The leaf nodes correspond to the convex subspaces created by the BSP-tree. A Boolean in a leaf node indicates whether the convex subspace is inside or outside the object.

At the University of Leiden, the object BSP-tree is used in the 3D graphics modelling system Hirasp (Teunissen and van den Bos 1988). Because of the spatial organization, hidden surfaces can be 'removed' in $O(n)$ time with n the number of polygons in the tree (Teunissen and van Oosterom 1988). The object BSP-tree is also well suited to perform the set operations (Thibault and Naylor 1987): union, difference and intersection, as used in constructive solid geometry (CSG) systems. The map overlay operation in a GIS (van Oosterom 1988a, 1988b) has strong relationships with these set operations.

3.1.3   The Multi-Object BSP-tree

We wish to exploit the spatial organization properties of the BSP-tree in a GIS. In a GIS we usually deal with 2D maps. The line segments of the original database are used to split the space in a recursive manner. By using data inherent in the problem to organize the space, we expect a better spatial organization, compared with the situation where fixed split lines are used, for example in the Quadtree. Maps always contain multiple objects, such as countries on the map of Europe. In order to deal with multiple objects, we have to modify the concept of the object BSP-tree previously discussed. Instead of a Boolean, the leaf nodes now contain an identification tag (a name). This identification tag tells to which object the convex subspace, represented by the leaf node, belongs. We call this type of BSP-tree the multi-object BSP-tree.

Fig. 3.3 [a] presents a 2D scene with two objects, triangle T with sides abc, and rectangle R with sides defg. The method divides the space in the convex subspaces of Fig. 3.3 [b]. The BSP-tree of Fig. 3.3 [c] is extended with explicit leaf nodes, each representing a convex part of the space. If

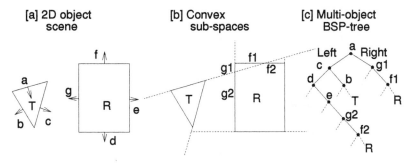

FIG. 3.3. The Building of a Multi-Object BSP-tree

a convex subspace corresponds to the 'outside' region, no label is drawn in the figure. If no more than one identification tag per leaf is allowed, only mutually exclusive objects can be stored in the multi-object BSP-tree, otherwise it would be possible also to deal with objects that overlap. A disadvantage of this BSP-tree is that the representation of one object is scattered over several leaves, e.g. rectangle R in Fig. 3.3. Fig. 3.4 shows the use of the BSP-tree for a simple polygonal map. As usual, the internal nodes contain the line segments and the leaf nodes represent convex parts of the space. The disadvantages of the BSP-tree become clear in this figure: it is sometimes necessary to split the line segments (edge f), and a polygon may be divided into two or more leaf nodes (region II into leaves 6, 7, and 9). The following list summarizes the properties of the multi-object BSP-tree:

- Each node in the tree corresponds to a convex subspace.
- Each internal node splits a convex subspace into two convex parts: left and right. Further down the tree, the convex subspaces become smaller. Each internal node contains one line segment.
- Each leaf node corresponds to a convex subspace which will not be split any further. A leaf node does not contain a line segment, but it does contain an object identification.

3.2 Basic Spatial Operations

This section will explain how the (multi-object) BSP-tree is used in implementing two spatial operations: the pick and the rectangle search.

3.2.1 The Pick Operation

Consider a system that displays a map on the screen. The user generates a point $P = (x, y)$ with an input device such as a mouse or tablet. He wants to know which object he has pointed at. To solve this problem we locate point P by descending the BSP-tree until a leaf node is reached. This leaf node contains the identification tag of an object. Descending the tree is quite simple: if, at an internal node, point P lies on the left side of the line segment, then the left branch is followed; otherwise the right branch is

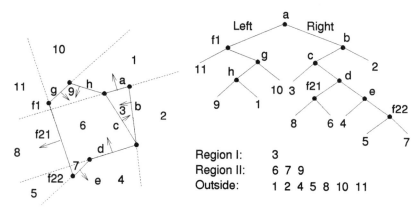

Region I: 3
Region II: 6 7 9
Outside: 1 2 4 5 8 10 11

FIG. 3.4. The BSP-tree of a Polygonal Map

followed. This strategy results in one straight path from the root to a leaf node. So, in case of a balanced tree with n internal nodes, the search takes $O(\log n)$ time; see Section 3.5. This is probably the best result one could hope for.

3.2.2   The Rectangle Search

In many applications the user wants to select all objects within a certain rectangle R. The rectangle search is also necessary during the display of (a part of) a map on a rectangular screen. Basically, the traversal of the tree is the same as in the pick operation. At an internal node, the left branch is followed if there is an overlap between rectangle R and the left subspace; the right branch is followed if there is an overlap between the right subspace and the rectangle R. If there is overlap with both subspaces, then both branches must be followed. A simple recursive function accomplishes this traversal. The operations are efficient because parts of the tree are skipped. In an unstructured collection of data, we would have to visit every item and test if we 'accept' this item based on its geometric properties. Using the BSP-tree, we do not have to examine the data that lie outside our region of interest.

3.3   Detail Levels

We need detail levels, as argued in Section 1.7, if we want to build usable interactive GISs. The detail levels must not introduce redundant data storage and must be combined with the spatial data structure. Not only must the geometric data be organized with detail levels, but the same applies to the related thematic data. However, we will focus our attention on the geometric data.

We first make an observation of the BSP-tree created with the function AddLine. A line segment inserted early on ends up in one of the top levels of the BSP-tree. A line segment inserted later on must first 'travel down'

[a] The global data   **[b] The detailed data**

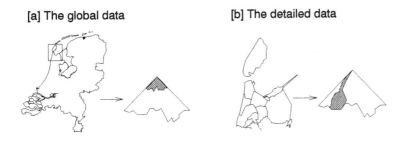

FIG. 3.5. The Place of Global and Detailed Data

the tree (and if necessary be split a few times) before it reaches the correct position on a lower level of the BSP-tree. We use this property to create a reactive BSP-tree. As far as I know, this reactive BSP-tree is the first reactive data structure ever presented (van Oosterom 1989). If the global data are inserted first in the BSP-tree, they will end up in the higher levels of the BSP-tree. Local data (detail) are added later, so they end up in the lower levels of the BSP-tree. Fig. 3.5 depicts this situation for a map of the Netherlands. The rectangle in the global map shows the position of the detailed map. The mountain-like object represents the entire BSP-tree and the grey region stands for the part of the BSP-tree that contains the data of the corresponding map.

I shall use the presentation of census data (see Subsection 1.5.2), involving the boundaries of administrative units, to illustrate the way the reactive BSP-tree functions. In the Netherlands there exist a number of hierarchical levels of administrative units, ranging from the municipalities (the lowest level) to the whole country (the highest level). We insert the boundaries of the administrative units in the BSP-tree, starting with the highest level, then the next highest level, and so on. During display of the map, the number of detail levels shown depends on the size of the selected region. The larger the region we want to display, the fewer detail levels will be shown. An heuristic rule for this is: the total amount of geometric data to be displayed should be constant, measured by the number of coordinates involved. Stated more formally: the pictorial information density must be constant.

The BSP-tree is traversed with an adapted 'rectangle search' algorithm, to display all objects in a certain region up to a certain detail level. The algorithm has to know where one detail level stops and the other begins. This can be achieved by extending the BSP-tree in one of the following manners.

1. The first approach adds to each node a label with the corresponding detail level. If, during the traversal of the BSP-tree, a detail level is reached that is lower than the one in which we are interested, we can skip this branch, because it contains only data of a lower level.

2. After inserting the global data (highest level) into the BSP-tree, the second approach adds special nodes, called level stop nodes, to the BSP-tree. The level stop nodes contain no splitting line segment and can be

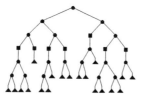

FIG. 3.6. The Reactive BSP-tree

compared with the leaf nodes of the multi-object BSP-tree (see Subsection 3.1.3). Then the next highest level is added to the BSP-tree, again followed by level stop nodes. This process is repeated for each detail level given. Fig. 3.6 shows a reactive BSP-tree with two detail levels.

A drawback of the reactive BSP-tree is that it supports only a part of the map generalization process (Shea and McMaster 1989). It removes unimportant lines but it draws an important line with the same number of definition points at every scale. As far as we know, there is no elegant solution to this problem in the context of the BSP-tree. It is possible to store a generalized version of a line at multiple detail levels in the same BSP-tree. However, the storage of the same line at multiple levels introduces unwanted redundancy. The generalized version of a line can be computed explicitly for every level with a line generalization algorithm, for instance with the Douglas–Peucker algorithm (1973).

3.4   An Application

In this section some additional uses of the reactive data structure in thematic mapping are described. I will expand the case of the previous section, to enable visualization of census data of administrative units. We shall see how a choropleth map and a prism map can be produced. Fig. 3.7 gives an example of these map types. In a choropleth map each region is assigned a colour dependent on the value of the thematic attribute in that region. In a prism map the theme is visualized by giving each region a certain height.

We use the reactive BSP-tree with level stop nodes to implement this application. A level stop node corresponds to a convex part of an administrative unit at that specific level. The identification of the administrative unit is stored in each level stop node. The census data are not stored in the BSP-tree, because the BSP-tree scatters objects (administrative units) over several leaves (see Section 3.1). The census data are available at each detail level.

3.4.1   Choropleth

After the user has decided which region and which census variable has to be displayed, the GIS determines the detail level. A choropleth map colours administrative units depending on the value of the census variable wanted. All we have to do to produce a choropleth map is to traverse the BSP-tree for the selected region and level. When we reach a level stop node of

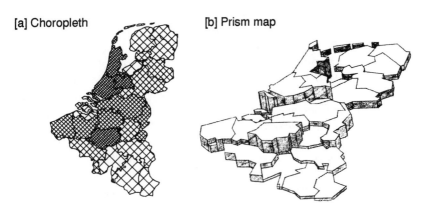

[a] Choropleth [b] Prism map

FIG. 3.7. Two Map Types for Thematic Mapping

the desired level, we know to which administrative unit the corresponding convex subspace belongs. The required census variable is retrieved and the convex subspace is filled with the correct colour.

The BSP-tree does not offer an explicit representation of the convex subspaces. This is solved by maintaining a temporary data structure during the traversal of the BSP-tree. This temporary data structure represents the (open) convex subspace that corresponds to the current node. Each time we take a step down the BSP-tree, the temporary data structure is updated. At depth k the temporary data structure represents a convex polygon with no more than k edges. So, a step downwards from level k to level $k + 1$ takes $O(k)$ time, the insertion of a new edge in a convex polygon with k edges. The steps upwards take no processing time, because the intermediate results are stored along the current path in the BSP-tree. In the case of a balanced BSP-tree, the height of the tree is $O(\log n)$ with n the number of line segments (nodes) in the BSP-tree. Summing all steps for the whole BSP-tree results in $O(n \log n)$ processing time. So, displaying the whole BSP-tree while colouring the convex subspaces takes $O(n \log n)$ time. It is possible to store the explicit representation of the convex subspace (or the reference to the original polygon) in the level stop node. This reduces the time taken to generate the choropleth of the whole map to $O(n)$, but increases the storage requirements. Generating only a part of the map takes $O(\log n + t)$ with t the number of convex regions that are found.

3.4.2 Prism Map

The prism map (Franklin and Lewis 1978) is an attractive map to look at and it offers the possibility to display an extra variable through the height of the prisms. A prism map is a set of 3D objects. Before the map is generated, the user has to indicate from which direction he wants to look at the prisms.

Basically, we produce the prism map in the same manner as the choropleth map. Instead of colouring the convex subspace, we lift it up to the desired height. If the convex subspace has k sides, then each side will result in a 3D rectangular polygon. Together with the top of the prism, this results in $(k+1)$ 3D polygons, which must be displayed. Before a polygon is displayed it is projected from 3D to 2D, in order to calculate the actual coordinates on the screen. A number of convex prisms form one prism on the map, as the corresponding convex subspaces together form the administrative unit. This means that the 'internal' sides of the prism need not be drawn. We can recognize the internal sides if we label those sides of the convex subspaces that are part of line segments.

The parts of the prisms that are not visible, because they are hidden by other prisms, must be removed. This 'hidden surface' problem is usually quite difficult and time-consuming to solve. However, if we change the way in which the BSP-tree is traversed slightly, the hidden surface problem is solved easily (in combination with the Painter's algorithm) (Newman and Sproull 1981). The adapted traversal does not cost any extra processing time and ensures that prisms farther away from the viewing point are drawn first. Therefore, parts of prisms are covered by other prisms that lie in front of them, because the prisms that lie in front are drawn later. This results in the 'virtual removal' of the hidden surfaces in the prism map. For more details on this topic see Teunissen and van Oosterom (1988).

Normally, the entire BSP-tree is traversed in $O(n)$ time, but we have to maintain the temporary data structure containing the explicit representation of the current convex subspace. So, we can produce a prism map of the whole scene in $O(n \log n)$. If explicit representations of the convex subspaces are stored, a prism map can be produced in linear time. This fast response stimulates the end-user to take other views of the data.

3.5   Balancing the BSP-tree

The arguments in the previous sections assume a balanced BSP-tree, but the algorithm in Fig. 3.2 will not necessarily generate a balanced BSP-tree. In fact, in some situations it is impossible to generate a balanced BSP-tree; see for example the 'convex scene' of Fig. 3.8. We can only solve this by inserting first some invisible auxiliary splitting line segments, for example a line with line segments a and b to the left and c and d to the right (not drawn in Fig. 3.8).

A balanced BSP-tree might result in a tree with more nodes, because of the splitting process. Sometimes, a slightly less balanced BSP-tree with fewer split line segments is preferred. This raises the question: What is the best BSP-tree for GISs? There is no simple answer to this question, but as long as both the measure in which the tree is out of balance and the number of split line segments remain within 'reasonable' bounds, it is my experience that the BSP-tree is well suited for GIS applications. The next two subsections describe several strategies for balancing BSP-trees in

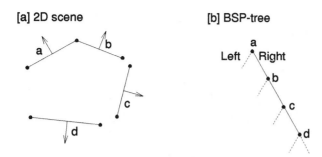

FIG. 3.8. Unbalanced BSP-tree

the static and the dynamic cases respectively. Dynamic balancing is not as important as in many other applications that use balanced trees, because the maps tend to be quite static in most GIS applications.

First, a general remark on balancing (Knuth 1973). There are two main criteria for balancing binary trees:

- Height balancing: the height of the left subtree may not differ more than a fixed number h from the height of the right subtree.
- Weight balancing: the number of nodes in the left subtree may not differ more than a fixed number w from the number of nodes in the right subtree.

As height balancing and weight balancing are strongly correlated, both will suit the needs of GISs in practice.

3.5.1   Static Balancing

In our implementation the line segments are, per detail level, inserted in the same order as they are stored in the original map file. In case of the map of the Netherlands, this results in a region-by-region insertion of the map data into the BSP-tree. For example, if the northern-most region is inserted first, then the paths that correspond to the area above the northern-most region will not grow when inserting the other regions. In this manner the BSP-tree gets out of balance. A simple solution is to insert the line segments, per detail level, in truly random order.

Another, more expensive, solution for balancing the BSP-tree is taken from Fuchs *et al.* (1983). A few potential roots for the tree are tried and the one that gives a balanced division is selected. Fuchs uses this solution in combination with the original (non-incremental) version for building a BSP-tree. Balancing the BSP-tree and minimizing the number of splits are two objectives that do not always agree. Thibault and Naylor (1987) describe some heuristics for evaluating the candidates.

A different approach is to insert first a few auxiliary split lines, which try to divide the space in a fair manner. The map data line segments are inserted after the auxiliary lines and end up in the proper regions. The auxiliary lines are marked as invisible. A disadvantage of the auxiliary

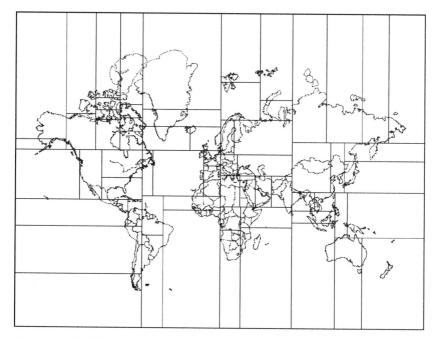

FIG. 3.9. A KD-tree with Large Bucket Size

lines is that they themselves may cause the split of line segments. However, experience has shown that this number of splits is relatively small. Sometimes the total number of nodes is even slightly reduced (though not significantly), in spite of the insertion of auxiliary split lines. Assume that the map data space is $\{(x, y)|0 \leq x \leq 1 \wedge 0 \leq y \leq 1\}$. We first insert the lines $x = \frac{1}{2}$ and $y = \frac{1}{2}$, then $x = \frac{1}{4}, x = \frac{3}{4}, y = \frac{1}{4}$ and $y = \frac{3}{4}$, and so on. Together, these auxiliary split lines form a coarse raster. Note that the order in which they are inserted is important. The insertion of 15 horizontal and 15 vertical lines reduced the (maximum) depth of a BSP-tree of the Netherlands (Map 1; see Section 3.6) from 80 to 35. The disadvantage of these auxiliary split lines is that they still result in unbalanced trees if the distribution of the map data is not uniform.

A more radical approach is first building a KD-tree (see Subsection 2.1.1), with large bucket size (e.g. 100–1,000). Because the KD-tree is suitable only for storing points, it is built from the end-points that define the line segments. Fig. 3.9 shows the KD-tree of a map that contains about 30,000 points. The split lines in the KD-tree are the auxiliary lines for the balanced BSP-tree, and the KD-tree is not used any more.

We could also use a generalized version of the KD-tree (see Fig. 3.10), which does not always split along one of the main axes. The BSP-tree is already suitable for storing split lines that have an arbitrary orientation. So it might be better to split along a line orthogonal to the best-fit line. The

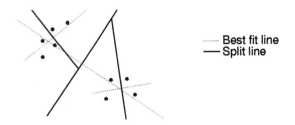

FIG. 3.10. The Generalized KD-tree

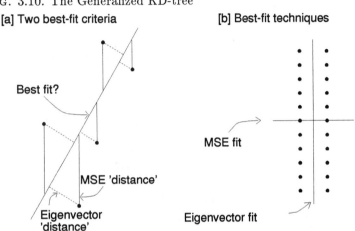

[a] Two best-fit criteria   [b] Best-fit techniques

FIG. 3.11. Eigenvector *v.* MSE Line Fitting

technique we used is called Eigenvector line fitting (Duda and Hart 1973). Fig. 3.11 compares eigenvector line fitting and the well-known statistical method mean square error (MSE) line fitting (Press *et al.* 1988). Eigenvector line fitting produces better fit lines because the Euclidean distance is used instead of the distance along the y-axis (see Fig. 3.11 [a]). Note that the term 'MSE line fitting' might be confusing, because not only the statistical (known as the MSE) line fitting method, but also the Eigenvector line fitting method minimizes the square error. The difference becomes clear in Fig. 3.11 [b]. Assume that the set of points for which an Eigenvector fit has to be calculated consists of $p_i = (x_i, y_i)$ for i from 1 to n. The general form of a line l that makes an angle α with the positive x-axis is: $x \sin \alpha - y \cos \alpha = c$. The Euclidean distance from point p_i to line l is: $|x_i \sin \alpha - y_i \cos \alpha - c|$. We minimize the function:

$$f(\alpha, c) = \sum_{i=1}^{n} (x_i \sin \alpha - y_i \cos \alpha - c)^2$$

With the means $\mu_x = \frac{1}{n} \sum_{i=1}^{n} x_i, \mu_y = \frac{1}{n} \sum_{i=1}^{n} y_i$, the variances $\sigma_x^2 = \frac{1}{n} \sum_{i=1}^{n} (x_i - \mu_x)^2, \sigma_y^2 = \frac{1}{n} \sum_{i=1}^{n} (y_i - \mu_y)^2$, and the covariance $\text{Cov}(x, y) = \frac{1}{n} \sum_{i=1}^{n} (x_i - \mu_x)(y_i - \mu_y)$ defined in the usual manner (Mood *et al.* 1974), we obtain α and c of the Eigenvector fit line:

$$\alpha = \tfrac{1}{2} \arctan \left(\frac{2\mathrm{Cov}(x,y)}{\sigma_x^2 - \sigma_y^2} \right), \quad c = \mu_x \cos\alpha + \mu_y \sin\alpha$$

for $\sigma_x^2 > \sigma_y^2$. If $\sigma_x^2 < \sigma_y^2$, then $\tfrac{1}{2}\pi$ must be added to α. If both variances are equal, then $\alpha = \pm\tfrac{1}{4}\pi$ depending on the sign of the covariance. The points are sorted according to the position of their projections on line l. All points up to the median are put in the left subspace and the others are put in the right subspace. This process is repeated for all subspaces until they contain fewer points than the specified bucket-size. This process results in a perfectly balanced tree for storing n points in $O(n(\log n)^2)$ time. The sorting causes some extra pre-processing time. If the set is split into two parts by using a split line through (μ_x, μ_y) and orthogonal to line l, then the building of the generalized KD-tree takes $O(n \log n)$ time. However, this does not necessarily result in a perfectly balanced tree.

Paterson and Yao (1989) prove that, if the original line segments are disjoint, then it is possible to build a BSP-tree with $O(n \log n)$ nodes and depth $O(\log n)$ using an algorithm requiring only $O(n \log n)$ time. They (recursively) use a horizontal auxiliary split line defined by the median value of the y-coordinate, and with the notion of 'free cuts' they prove their theorem. Their results can easily be generalized to line segments representing a map and touching each other only at start- and end-points. Better partitions may be achieved by using more general auxiliary split lines, e.g. lines similar to the ones of the generalized KD-tree. It is doubtful whether this BSP-tree can be built in $O(n \log n)$ time. Among other difficulties, the detection of free cuts will become harder.

3.5.2   Dynamic Balancing

The emphasis in this subsection is on inserting line segments in a BSP-tree, while keeping it balanced. Deleting and modifying line segments have less importance, because they occur less frequently in GIS applications. Even without considering the balance of the BSP-tree, really deleting a line segment can be very difficult. If the line segment is the root of a (sub) BSP-tree, then the replacement of this root by another line segment affects the whole subtree in a drastic manner. This is not a problem in the case of an empty or very small subtree but otherwise could require the complete rebuilding of the subtree. A deletion can be simulated by making the line segment invisible, as with an auxiliary line. It will be clear that this is not a practical solution when the number of deletes and changes is relatively large compared with the actual number of line segments.

To enable dynamic balancing, the nodes in a BSP-tree need to be extended with information about their balancing status. This is a single parameter that contains for example the value of the expression #NodesLeft − #NodesRight. The dynamic insertion of a line segment starts in the same manner as in the normal situation, i.e. with the function `AddLine` (see Fig. 3.2). During the insertion the balance status of the visited nodes has to be updated. However, because of this insertion it is possible that nodes on the path from the root to the new leaf get out of balance. Note that, in case of a split line segment, there are several leaves that correspond to the new line segment. There may be multiple paths from the root that have to be considered during the restoration of the balance and, as a consequence, the weight associated with subtrees may increase with more than one.

In order to restore the balance, a subtree that corresponds to an unbalanced node has to be reorganized. This is done for each node that is out of balance. A solution might be to perform a complete rebuild of a subtree based on (exhaustive) search for a good root in the subtree. This could be done in a way comparable to the method Fuchs (1983) describes to balance a BSP-tree in the static case. This method is not only very time-consuming, but in case of 'convex scenes' is even impossible, as explained in the introduction of this section. Therefore, I decided to use a new technique.

The root of the new subtree is an auxiliary line and it is made in a similar way to how a line in the generalized KD-tree is created. If, during the calculation of this auxiliary line, a point that is an end-point of more than one line segments is also counted more than once, then the number of line segments to the left and to the right are equal. In order to preserve as much as possible of the (balanced) structure of the old subtree, we should try to move parts as large as possible from the old to the new subtree. To simplify the test of whether a part fits in the new subtree, a circle is stored in each internal node of the BSP-tree. The centre lies halfway along the line segment and the radius is the smallest value such that all line segments of the subtree lie within the circle. The computational speed is further increased by storing the square of the radius instead of the radius itself. This circle together with the BSP-tree structure of the new subtree makes it easier to move parts of the old subtree.

3.6 Practical Results

In this section the first results of this implementation are presented. Note that this is only a prototype GIS and not all functions are present yet. The prototype is a 'main memory implementation'; that is, the complete map is stored in a data structure of a running program. Especially for large data sets, it would be useful to perform a redesign of the prototype, in order to minimize the number of disk accesses during a tree search. The most obvious strategy is to take a group of nodes (a small subtree) of the BSP-tree and store them on one disk-page. If each disk-page is considered to be one node in the 'external implementation', then the resulting structure is a multi-way tree. Probably, this structure will resemble the Cell tree described by Günther (Günther 1988, Günther and Bilmes 1988, 1991); see Subsection 2.1.8.

Fuchs *et al.* (1980) show that n line segments may result in the worst-case in $O(n^2)$ line segments in the BSP-tree, because of the splitting pro-

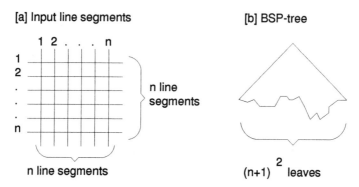

FIG. 3.12. Worst-Case Situation for Line Segments: A Line Raster

cess. This happens when the line segments are long, relative to each other, and have inconvenient orientations and positions. The insertion of one (long) line segment results in many new leaves in the BSP-tree (see Fig. 3.12). In a balanced tree this is no problem for the query time: $Q(n) = O(\log n^2) = O(2\log n) = O(\log n)$. However, it results in an enormous storage requirement: $S(n) = O(n^2)$. This is unacceptable in the case of GISs in which n is typically very large, e.g. 10,000–100,000. As already noted in Subsection 3.5.1, in case of disjoint line segments the storage requirement is $S(n) = O(n\log n)$.

How will the BSP-tree behave when we insert very large amounts of irregular geometric data? In contrast to the worst-case, we expect the number of splits in the practical GIS situation to be far less because the line segments are relatively short. We are interested in the size and the performance of BSP-trees built with real map data. In table 3.1, Map 1 is the map of the Netherlands as drawn in Fig. 1.8 on p. 14. Map 2 contains the data from World Data Bank I (WDB I). The area and line features from DLMS DFAD (DMA 1977) are used in Map 3. The latitude ranges from 52°12′ to 52°24′ and the longitude from 5°30′ and 6°00′; this is the region near Harderwijk in the Netherlands. These three maps are from completely different sources, but they produce very similar test results. The table shows some of the key figures when no measures for balancing have been taken.

The expansion factor is defined by $f = L/I$ where L is the number of leaves in the tree and I is the number of inserted line segments. The theoretic minimum depth is defined by $d_{th} = \lceil {}^2\log L\rceil$. The maximum depth d_{max} and the average depth d_{avg} are measured values. Degenerated line segments are line segments in the original data set with the start-point equal to the end-point or at least within a distance smaller than a relative accuracy epsilon, as used by our program. The number of degenerated line segments is indicated by D. There is a simple relationship between the number of leaves in the BSP-tree and the number of inserted, degenerated, and split lines: $L = I - D + S + 1$. This is due to the property of binary

Table 3.1. Key Figures of Some Actual BSP-trees (without Balancing)

	MAP 1 NETHERLANDS	MAP 2 WDB I	MAP 3 DLMS
Inserted line segments (I)	13,350	108,966	5,456
Degenerated line segments (D)	3	50	4
Split line segments (S)	6,313	48,173	2,586
Leaves (L)	19,661	157,090	8,039
Expansion factor (f)	1.47	1.44	1.47
Maximum depth (d_{max})	80	234	60
Average depth (d_{avg})	27.7	63.2	30.5
Theoretic minimum depth (d_{th})	15	18	13

Table 3.2. Balancing BSP-trees with a Simple Raster (Map 1 – the Netherlands; 13,530 inserted line segments)

	NO RASTER	DUMB 16×16	DUMB 32×32	SMART 16×16	SMART 32×32
Split line segments (S)	6,313	6,260	6,831	6,119	6,318
Leaves (L)	19,661	19,642	20,243	19,637	20,347
Expansion factor (f)	1.47	1.46	1.51	1.45	1.45
Maximum depth (d_{max})	80	52	73	35	30
Average depth (d_{avg})	27.7	28.4	40.3	16.9	14.7

trees that 'the number of external nodes is one more than the number of internal nodes' and is corrected for split line segments and degenerated line segments. The key figures for a BSP-tree that is balanced by a coarse raster can be found in Table 3.2. In addition to the 'smart' method for inserting the raster described in Subsection 3.5.1, the table also gives the results for a more 'dumb' method. The 'dumb' method inserts the line segments from the raster from left to right and from top to bottom, before inserting the map data. The BSP-trees created by the 'dumb' method have larger average and maximum depth compared with the ones created by the 'smart' method. Note that the number of split line segments does not have to increase because of the extra line segments of the raster.

Fig. 3.13 shows the number of split line segments as a function of inserted line segments for the data from WDB I. One might expect that the more line segments have already been inserted in the BSP-tree, the bigger the change that a new line segment has to be split. However, this is not true. The straight line in Fig. 3.13 means that the chance that a new line segment has to be split is independent of the number of already inserted line segments.

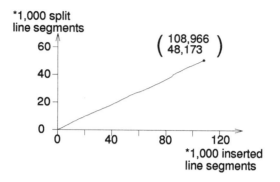

FIG. 3.13. Split Lines as Function of Inserted Lines

This is a remarkable result, because it implies that BSP-trees of real maps have linear $O(n)$ storage space complexity in contrast to the worst-case quadratic $O(n^2)$ and the $O(n \log n)$ for disjoint line segments. The constant associated with this $O(n)$ storage space complexity is modest and stable, somewhere between 1.4 and 1.5. An intuitive explanation for this is that the line segments have 'point-like' characteristics because they are small compared with the entire map. When the line segments reach their final position, they gradually get back the line characteristics. Another approach to analysing this $O(n)$ storage space complexity is the development of a statistical model.

3.7 Discussion

The data structure presented is one of the few that combines the two difficult requirements: spatial organization and detail levels. Because of its generality, it enables incorporation of other spatial organization techniques into the BSP-tree, e.g. the raster structure, the Quadtree (Samet 1989), or the KD-tree (Bentley 1975). A surprising result of this implementation is that BSP-trees of real maps seem to have no more than about $1.5 * n$ nodes instead of the worst-case $O(n^2)$ nodes with n the number of line segments in the map.

I am aware of the fact that the reactive BSP-tree is far from perfect, but I hope that it serves as a source of inspiration to generate more ideas. One important guideline derived from my work on the BSP-tree is that, in a spatial organization tree, the more important objects must be stored in the higher levels of the tree. Using the insertion order as done in Section 3.3 is not a very elegant method of achieving this because the solution is limited to static data. A reactive data structure (van Oosterom 1989) need not be based on a BSP-tree; other solutions are possible. In Chapter 6, a few other types of reactive data structures are presented.

4 Orientation-Insensitive Indexing Techniques

Axes-parallel rectangular and grid methods for storing geometric objects are particularly suited to deal with spatial search queries that select everything within an axes-parallel rectangle. In general, axes-parallel methods are appropriate when dealing with sets of data that are oriented parallel to the main axes, as for example in very large-scale integration (VLSI) design. However, in many applications the data and the query regions are more irregular, and in those cases the performance of these methods suffers. Examples of such queries are: 'Find the n-nearest-neighbours'; 'Select everything within a polygon, a circle, or a rotated rectangle'. The last one is useful for example in a navigation system with heading-up display which has the task of continuously showing everything within an arbitrary oriented rectangle.

Two new spatial organization techniques are proposed for storing and retrieving large sets of geometric objects: the KD2B-tree and the Sphere-tree. The main characteristic of these two techniques is that they are orientation-insensitive. This means that, in contrast to nearly all other known spatial organizations, the expected spatial query time is independent of the orientation of the data and the query regions.

My colleagues and I tested our indexing techniques with large map data sets and compared the results with the the well-known R-tree. Among other data sets, we used World Data Bank II and a large random data set for our performance tests. Our indexing techniques outperformed the R-tree in the case of overlap queries with circles and convex polygons. We have tried to design spatial organization techniques that, apart from being orientation-insensitive, have the following general properties:

- *Geometric object-oriented:* Objects are not necessarily zero-sized; i.e., they are not limited to points and may have spatial extent. However, they must be treated as one unit; that is, they may not be divided into parts as a result of the spatial organization.
- *Dynamic:* It must be possible to insert and delete objects without a global restructuring of the spatial organization in order to guarantee efficient memory usage and short query times.
- *Integrated external disk storage scheme:* Very large sets of geometric objects must be stored in a persistent manner. This is in contrast to spatial data structures that reside in main memory only.
- *k-dimensional:* Though this is not required in our main application area, GIS, generalization to k-dimensional space is desirable. Unless

indicated otherwise, we will consider the 2D case in the remainder of this chapter.

A few orientation-insensitive search structures are known, for example the BSP-tree (Fuchs *et al.* 1980) and the Cell tree (Günther and Bilmes 1991). A common drawback of these two methods is that they are not geometrically object-oriented; that is, they sometimes subdivide an object into parts and store these parts separately. A related problem is that their dynamic properties are limited. Two new spatial indexing methods will be presented: the KD2B-tree and the Sphere-tree, described in Sections 4.1 and 4.2, respectively. Both index structures use the minimal bounding sphere of an object. Some advantages of using spheres instead of rectangles are: orientation-insensitivity, decreased storage space requirement (especially for higher-dimensional index structures), and in some cases reduced area coverage by the bounding spheres as compared with that of bounding rectangles. However, this last advantage depends heavily on the shape of the objects. For example, practice showed that in the case of 'polyline' objects the coverage of the rectangles was usually less than the coverage of spheres. This experience is supported by a statistical analysis in Section 4.4.

In order to evaluate the performance aspects of the two new indexing methods, a comparison is made with the well-known R-tree (Guttman 1984). Several digital maps are used for this purpose. Information about the development and test environment can be found in Section 4.3, which covers the topics hard- and software, map data sets, query types, object-oriented modelling of index structures, and some implementation details. Section 4.4 presents the performance test results together with some interpretations. Finally, conclusions and further research are described in Section 4.5.

4.1   The Design of the KD2B-tree

The design of the KD2B-tree (Claassen 1989) finds its roots in the KD-tree (Bentley 1975) with 'split' lines that may have arbitrary angles, the so-called generalized KD-tree (van Oosterom and van den Bos 1989*a*, 1989*b*); see also Sections 2.1 and 3.5. The advantages of arbitrary angles are that the partitioning of the space is more adaptable to the data distribution and that the resulting structure is orientation-insensitive. Fig. 4.1 shows the difference between the KD-tree and the generalized KD-tree. Before describing the KD2B-tree itself, I shall first describe its predecessor, the KD2-tree: a main memory index structure. The requirement to treat non-zero-sized objects as a unity led to the development of the KD2-tree. Subsequently, this tree was adapted for external storage in a manner similar to the KDB-tree (Robinson 1981). Briefly summarized, the KD-tree and the KDB-tree are known structures; the KD2-tree and the KD2B-tree are new structures.

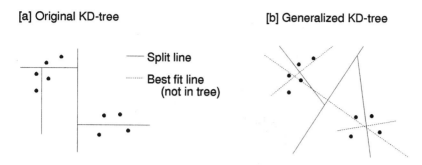

[a] Original KD-tree   [b] Generalized KD-tree

—— Split line

······ Best fit line
(not in tree)

FIG. 4.1. The Difference between a KD-tree and a Generalized KD-tree

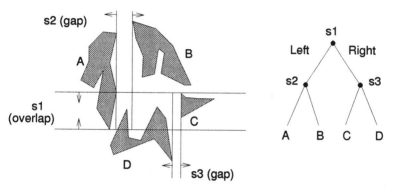

FIG. 4.2. The Principle of the KD2-tree

4.1.1 The KD2-tree

The KD-tree can only handle spatial objects of zero size, that is points. In many GIS applications a point location is not enough for a good representation of a spatial object. For example, objects like rivers, railroads, counties, etc., occupy regions with a 2D spatial extent. Some modifications are needed to transform the generalized KD-tree into a new version of the KD-tree, the KD2-tree, which can handle spatial objects of non-zero size. The main changes are:

- to use two parallel split lines per internal node, because of the possible overlap between objects that have to be stored in different subtrees (hence the name KD2-tree) (see Fig. 4.2);
- to treat an object as a unit by using, e.g., a minimal bounding rectangle (MBR) or sphere (MBS) and a reference to the object.

The KD2-tree is a statically balanced tree, so all the leaf nodes are at the same level. A balanced tree has the advantage of having a better worst-case search time compared with a tree that is not balanced. A leaf node contains index record entries of the form: (MBS, ObjectId) where MBS is a minimal bounding sphere, in the 2D case a circle, completely containing the geometric object, and ObjectId refers to the object data. In

Subsection 4.2.1 I will give an algorithm to calculate the MBS. I use the MBS instead of the MBR because the split lines of the generalized KD-tree have arbitrary angles, for which spheres are better suited. An internal node contains an entry of the form: (LeftLine, LeftId, RightLine, RightId) where LeftLine and RightLine are two parallel split lines each defining an associated subspace. LeftId refers to the node which corresponds to the 'left' subspace. The definition of the RightId is analogous. A line l with normal vector $\vec{n}$ is described by: $\{\vec{x}|(\vec{n} \cdot \vec{x}) = c\}$. Point $\vec{p}$ lies to the left of line l if $(\vec{n} \cdot \vec{p}) < c$; point $\vec{p}$ lies to the right if $(\vec{n} \cdot \vec{p}) > c$, otherwise point $\vec{p}$ lies on l.

The algorithm for constructing the KD2-tree starts with the calculation of the 'best-fit' line approximating the centre-points of the current set of objects (van Oosterom and van den Bos 1989a, 1989b). The left and right split line are orthogonal to the best-fit line. The set is divided into a left and a right subset. The 'left' split line is shifted, such that it defines the smallest subspace that completely covers all the objects of the left subset. In the same manner, the 'right' split line is determined. This process continues recursively until all spatial objects in the current space fit into one leaf node. Note that there may be either a gap or an overlap between the two regions; see Fig. 4.2. Basically, the search operations for the KD2-tree are similar to those for the KD-tree. That is, when the search region overlaps the left and/or the right subspace, the search is continued recursively in the subtree that corresponds to the overlapping subspace until the leaf node is reached.

Besides being able to store objects with a spatial extent, another advantage of the KD2-tree is the possibility of representing gaps, i.e. empty regions between sets of objects, which speeds up the search process. This will be useful in the case of relatively small objects. The KD2-tree is not dynamic: insertion or deletion of a spatial object is not possible without disturbing the balance of the tree. The KD2-tree has to be built up in main memory each time the application program is started, so it is not a persistent structure. In the next subsection, the KD2-tree is modified into a new structure that is dynamic and persistent. This structure is called the KD2B-tree, which has a similar relationship to the KD2-tree as the KDB-tree (Robinson 1981) has to the KD-tree.

4.1.2   The KD2B-tree

The KD2B-tree is a height-balanced tree with index records in its leaf nodes containing pointers to data objects. The nodes of the tree correspond to disk pages. The KD2B-tree is dynamic; insertions and deletions can be interchanged with searches and no periodic reorganization is required. The leaf nodes are the same as the leaf nodes of the KD2-tree, except for the fact that the leaf nodes of the KD2B-tree are stored externally on disk. An internal node contains a tree structure consisting of a small subset of internal nodes from the corresponding KD2-tree. Fig. 4.3 shows the conceptual structure: big 'clouds' represent the internal nodes, small

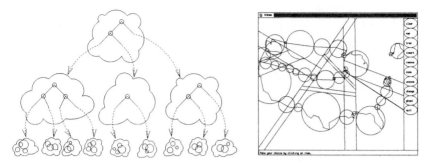

FIG. 4.3. The Conceptual Structure of the KD2B-tree

'clouds' represent the leaf nodes of the KD2B-tree. Internal nodes contain a number of index record entries each of which can be compared with one internal node of the KD2-tree, but now LeftId and RightId may be either pointers to entries within this node, or pointers to other nodes in the KD2B-tree (indicated with dashed lines in Fig. 4.3). A KD2B-tree has the following properties:

1. A leaf node has at most M_l entries.
2. A leaf node has at least $m_l(\leq \lceil M_l/2 \rceil)$ entries, unless it is the root which has at least one entry.
3. An internal node has at most M_i entries.
4. An internal node has at least one entry referring to at least one node, unless it is the root which has at least one entry referring to two nodes.
5. All leaf nodes are at the same level.

By choosing the proper values for the parameters M_l and M_i, the nodes of the KD2B-tree can be made to fit into one disk page.

4.1.3 The Operations

This subsection describes the Search, Insert, and Delete operations on the KD2B-tree. The region search is similar to the region search of the KD2-tree. The recursive Search algorithm is quite straightforward:

1. If the node is a leaf node, select all entries that have their MBS overlapping with the search region.
2. If the node is an internal node, invoke Search recursively for all subspaces that have overlap with the search region.

The insertion of a single entry can be used for building up the structure and for inserting new spatial objects. The algorithm for the Insert operation is:

1. Traverse the tree to find the best leaf by recursively choosing the most appropriate subspace until a leaf is reached. While moving down, the split lines in the nodes encountered on the path from the root have to be adjusted.
2. Add the entry to the selected leaf node.

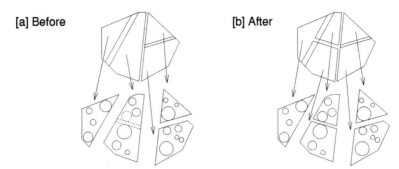

FIG. 4.4. Splitting a Leaf Node in the KD2B-tree

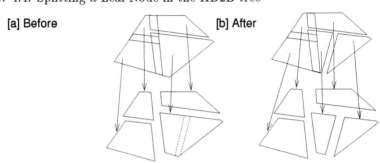

FIG. 4.5. Splitting an Internal Node in the KD2B-tree

3. If an overflow occurs, split the leaf into a left and a right leaf, and add the split lines to the parent (internal node); see Fig. 4.4. If overflow occurs while adjusting the parent, then also split the parent itself and adjust its parent; see Fig. 4.5. Repeat this process on the next higher level until there is no more overflow. If the root is reached and there is still overflow, create a new root.

The way nodes are split is important, because this affects the search time. There is quite a difference between the splitting of leaf nodes and that of internal nodes. The splitting of leaf nodes is similar to the division of a set of spatial objects into two subsets of almost equal size (i.e. differing by at most one), as described for the KD2-tree (see Fig. 4.4). Another method could be to minimize the overlap (to maximize the gap) between the left and right subspaces, subject to the constraint that the number of entries per subset has to be at least m_l. The splitting of an internal node is implemented by removing the root of the internal tree structure. The two children of this root become new roots. Each of these roots with corresponding tree is stored in a separate new node. The old root is added to the parent (see Fig. 4.5). Note that for clarity no entries, i.e. double split lines, are shown on the second level in Fig. 4.5. In this way the properties of the KD2B-tree remain valid, because the number of entries in the new nodes is at least one less compared with the original node, which had overflow. The deletion of existing spatial objects is done by the Delete algorithm which uses the Insert operation:

1. Find the leaf containing the entry, using its MBS. If the entry is not found, give an error message and stop, or else remove the entry from this leaf.
2. If underflow occurs, save all entries of the under-full leaf in a temporary structure and adjust its parent by removing the pointer. If the parent also becomes under-full, i.e. empty, remove this parent and adjust its parent in the same manner. If necessary, repeat this process until the root is reached.
3. Adjust all split lines corresponding to the regions to which the deleted object belonged.
4. If underflow has occurred, reinsert all saved entries of the removed leaf by using the Insert operation.

4.1.4 The Packed KD2B-tree

A drawback of the KD2B-tree is that there is no lower bound for the number of entries in an internal node. This may result in poor memory utilization and a less than optimal search structure. It is very hard to guarantee a higher number of entries per internal node in a dynamic situation. However, most nodes in the KD2B-tree are leaf nodes, and these nodes are guaranteed to be occupied by at least m_l entries. So the overall memory usage is acceptable. An aspect of the KD2B-tree that may be further improved is the creation of an initial tree. When the Insert algorithm is used for building the KD2B-tree, this may result in a poor spatial subdivision, because the objects that are inserted first are not representative for the whole set of objects; however, they determine the major split lines. A packing technique can be used for creating a better initial KD2B-tree, with two important improvements. First, it has a better spatial structure: parallel split lines are better positioned, because the best-fit lines are based on all centre-points of the set of objects contained in the corresponding subspace. Also, each internal node contains an internal balanced tree structure. Second, the packed KD2B-tree has a more efficient memory usage: leaf nodes are filled as fully as possible.

Conceptually, the creation of a packed KD2B-tree is the same as the construction of the KD2-tree with bucket size M_l. This would result in a balanced binary tree. However, this tree has to be divided into small balanced subtrees, each of which has to be stored in one internal node of the (multi-way) KD2B-tree.

In practice, the building of a packed KD2B-tree starts with an initialization phase and this is followed by a call to the recursive function BuildPack. In the initialization phase all n objects are inserted into the first $\lceil n/M_l \rceil$ nodes without creating an internal structure. These nodes will become the leaf nodes of the packed KD2B-tree. The height of a balanced binary tree 'on top of' all $\lceil n/M_l \rceil$ leaf nodes would be: $h_t = \lceil {}^2\log(\lceil n/M_l \rceil) \rceil$. The maximum height of a subtree that can be stored in one internal node of the

KD2B-tree is: $h_i = \lfloor 2 \log M_i \rfloor$. The height of the packed KD2B-tree will be: $h = \lceil h_t/h_i \rceil + 1$. In order to produce a compact balanced KD2B-tree, all the internal nodes other than the root and the internal nodes just above the leaf nodes have to be filled with a perfectly balanced subtree of height h_i. The internal nodes just above the leaf nodes are filled with subtrees of the same height, but not necessarily perfect. The root node of the KD2B-tree is filled with the perfectly balanced top of height: $h_t - ((h - 2)h_i) \leq h_i$.

The BuildPack algorithm must be invoked with a (sub)set of leaf nodes as input. It returns a pointer to the created internal node that contains the binary subtree corresponding to the set of input leaf nodes. To start the pack process, BuildPack is invoked with the whole set of initial leaves. The return value is the root of the KD2B-tree. As long as BuildPack has not reached the leaf nodes, it creates a new internal node and fills it with the predetermined maximum number of entries ($M_p = 2^{h_i} - 1$). Each entry is created in the following manner:

1. Determine the direction of the parallel split lines by first calculating the best-fit line through all centre-points of the current set of objects.
2. Sort along the best-fit line and split the set into a left and right subset (in the leaf nodes).
3. Determine the position of the split lines.
4. Store the parallel split lines in an unoccupied entry of the internal node and connect this entry with its parent.

If BuildPack has not created an internal node just above the leaves, it calls itself for each of the $M_p + 1$ subsets of leave nodes. After each invocation the new internal node of the KD2B-tree is connected.

The time complexity of this algorithm is dominated by the sorting of the entries. At the top level of the binary tree this takes $O(n \log n)$; at the second level it takes $O(2^{-1}n \log(2^{-1}n))$ and has to be done twice; at the ith level it takes $O(2^{1-i}n \log(2^{1-i}n))$ and has to be done 2^{i-1} times. As there are at most $\lceil 2 \log n \rceil$ levels, this results in:

$$\sum_{i=1}^{\lceil 2 \log n \rceil} 2^{i-1} \cdot 2^{1-i}n \log(2^{1-i}n) = O(n \log n \log n)$$

A packed KD2B-tree can be updated without losing the efficient spatial structure, as long as the number of updates is relatively small compared with the size of the whole data set (e.g. $n/4$), as in GIS applications. It will usually take quite a large number of insertions before the height of the tree increases, because each internal node contains $M_i - M_p$ free entries.

4.2   The Sphere-tree

This section will describe an orientation-insensitive index structure of which the nodes are filled at least for 50 per cent with entries. The first step in the design of the Sphere-tree was the use of bounding spheres to make the BSP-tree more dynamic (Fuchs *et al.* 1980, van Oosterom 1989). The incorporation of the external memory requirement resulted in a structure with 'external' nodes that each contain a number of split lines (similar to the KD2B-tree). Finally, we decided to get rid of the split lines, as used in the BSP-tree and the KD2B-tree, because the size of one external node is relatively small and the use of split lines is not necessary. This resulted in a structure that bore little resemblance to the original BSP-tree. It was interesting to see that the Sphere-tree developed into a structure similar to the R-tree (Guttman 1984) with MBSs instead of MBRs. Fig. 4.6 shows the R-tree and the Sphere-tree of the same map. Besides being orientation-insensitive, the Sphere-tree has the advantage over the R-tree in that it requires less storage space. In k-dimensional space a MBR requires two points, i.e. $2k$ floating point numbers, and a MBS requires only one point and a radius, i.e. $k + 1$ floating point numbers.

As the operations on the Sphere-tree are very similar to those on the R-tree, they are not described here. I will explain the algorithm used to calculate the MBS of an object (polygon or polyline) and also to calculate the MBS of a set of spheres, because this algorithm is more complicated than the similar calculation of the MBR. The MBSs of geometric objects end up in the leaf nodes of the Sphere-tree. The calculation of the MBS of MBSs is required for the internal nodes of the Sphere-tree.

4.2.1   The Minimal Bounding Sphere

The problem of finding the minimal bounding circle of a set of points in the plane dates back to 1857 when it was first posed by Sylvester (Sylvester 1857). In order to calculate the MBS of a polygon or polyline, it is sufficient to calculate the MBS of the set of points defined by their vertices. The algorithm to calculate the MBS is based on the one described by Elzinga and Hearn (1972). Conceptually this algorithm works in any dimension, is quite simple, and is the same for points (see Fig. 4.7 [a]) and spheres (see Figs. 4.7 [b] and 4.7 [c]). I shall refer only to an input data set consisting of spheres. According to Preparata and Shamos (1985), this algorithm uses $O(n^2)$ time in the worst-case with n the number of spheres in the input set. Our iterative algorithm to calculate the MBS of spheres in k dimensions is:

1. Select $k + 2$ spheres from the input set, e.g. the first $k + 2$.
2. Determine the MBS of the selected set and mark the sphere that was not used.
3. If the MBS of the selected set covers all spheres from the input set, then the solution is found. Otherwise, replace the unused sphere from the selected set with the sphere from the input set that is the farthest outside and go to step 2.

I made two modifications to the algorithm of Elzinga and Hearn. The first one makes it more efficient by taking the farthest sphere outside the MBS instead of the first sphere (in step 3). The second modification makes incremental use possible. This is done by allowing the input data set to

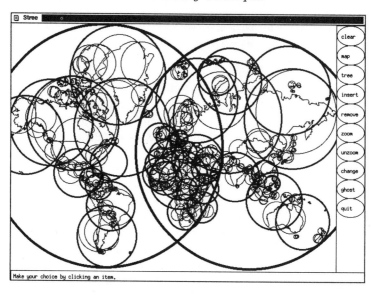

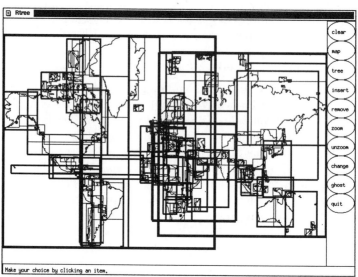

FIG. 4.6. The Sphere-tree Compared with the R-tree

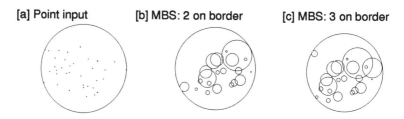

FIG. 4.7. The Construction of the Minimal Bounding Sphere

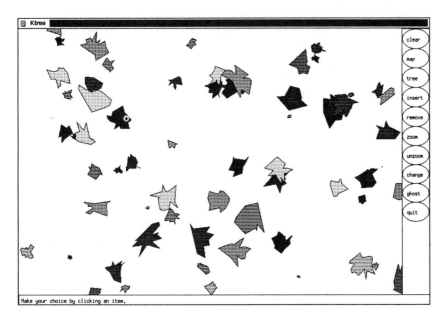

FIG. 4.8. A Small Part of the Random Data Set Art

be changed dynamically with 'Add' and 'Delete' sphere operations. In practice, the algorithm runs in linear time.

Step 2 of our algorithm will be illustrated in the two-dimensional case, in which the selected set contains four circles (2D spheres). There are three ($= k + 1$) possible types of solutions: the MBS is defined by one circle (one circle covers the others); the MBS is defined by two circles (see Fig. 4.7 [b]); or the MBS is defined by three circles (see Figs. 4.7 [a] and 4.7 [c]). This situation corresponds to the classic 'problem of Apollonius'. There is always at least one circle that is not used, even if four circles lie on the MBS. This is the one that is marked 'unused'. The proof that this algorithm is mathematically correct can be found in Elzinga and Hearn (1972). The algorithm is sensitive to the finite precision of floating point numbers. When no measures are taken to prevent this problem, the algorithm may loop for ever when $k + 3$ spheres are found to lie on the MBS and the same sphere is removed and inserted over and over again.

An alternative to our approach might be to use the algorithm of Megiddo (1983), which calculates the MBS of a set of points in linear time $O(n)$, based on linear programming techniques (Press *et al.* 1988). However, Megiddo's algorithm seems more difficult to implement and to adapt to our situation (MBS of MBSs, incremental). Another alternative is the algorithm described by Skyum (1990), which runs in $O(n \log n)$ and is quite simple.

Anyway, the calculation of MBSs is not done during the region search operations. So only 'Delete' and 'Insert' operations would be influenced.

Thus, the query time, one of the most important characteristics of the structure, would not improve.

4.3   The Development and Test Environment

The KD2B-tree and Sphere-tree indexing methods have been implemented in GNU C++ on a Sun 3/60 workstation (68020 processor, 68881 floating point coprocessor, and 8 Mb main memory) under SunOS 4.0, a Unix variant. The workstation has a local SCSI disk (MiniScribe 338 Mb, 16 ms average seek time). The R-tree has been implemented for comparison, because it is well-known and it also satisfies all the requirements listed in the introduction of this chapter, except for being orientation-insensitive. Several map data sets and random data sets were used for testing the index structures. These data sets contain three different types of geometric objects: points, polylines, and polygons. The largest map data set, World Data Bank II (WDB II) (Gorny and Carter 1987), contains more than 30,000 objects defined by more than 5,000,000 points. The size of the original ASCII data set is about 112 Mb. The total area covered by the MBRs and the MBSs is 37 and 76 per cent of the object data space, respectively. There are large gaps (e.g. oceans), but there are also many cases of overlap. The random data set, called Art, consists of 100,000 polygons created by a random generator (see Fig. 4.8). In Art the total area covered by the MBRs and the MBSs is nearly equal, 25 and 26 per cent of the object data space, respectively.

Four different types of queries were tested: overlap with point-, rectangle-, sphere-, and convex polygon-query objects. For each of the latter three types, four different query object sizes were tested, covering 0.01, 0.1, 1, and 10 per cent of the object data space. This results in a total of 13 query sets. The n-nearest-neighbour query was not implemented, because the results were expected to be similar to the sphere query. For each set, 2,000 uniformly distributed query regions were generated and stored in a file. This implies that each index structure was tested with the same queries. Actually, the query rectangles are squares, and the query convex polygons are squares rotated by $\pi/4$.

4.3.1   Object-Oriented Modelling of Index Structures

The object-oriented paradigm offers a good development environment. The functionality of the spatial search class (or object type) is defined by identifying the methods that are externally available. This interface is the same for all spatial access methods that were implemented. It turned out that the most generally usable class has the form of a spatial indexing method. The geometric objects themselves are not stored in the search structure, but only the references to these objects. The summarized list of available methods is: `NewIndex`, `OpenIndex`, `AddObject`, `DeleteObject`, `Search*` (several variants), `CloseIndex`. The R-tree, the KD2B-tree, and the Sphere-tree are 'software ICs' implemented as C++ classes and may easily be used by other programs. Some other classes of our implementation that are worth mentioning, are Node, Geometry, and Graphix. The class Node holds one node of the index structure and offers the methods AddEntry, DeleteEntry, and Clear. Geometry provides some basic geometric operations, e.g. overlap functions, and calculation of MBS and MBR. All index methods use the Geometry object and are therefore based on the same implementation of the geometric operators, which makes the comparison fair. The class Graphix is a simple object-oriented interface to X/11 with operations such as: SetTransformation, DrawPolygon, DefineMenu, and PutMessage. This class is used not by the indexing methods, but by an application program, which we used to perform some visual functionality tests.

4.3.2 Some Implementation Details

We use shared mapped memory (virtual files) to store the nodes of the index structure. A file is mapped directly by the Unix OS on the address space of a process. A major advantage is that there is no difference between the 'stored' object instances and their 'running' counterpart, at least not at the level of the index structures. At the OS level there is a difference and this is the same as the difference between virtual memory pages that are in main-memory and the ones that are swapped on disk. We compared this implementation with our previous one that did not use mapped memory, but used the Unix OS system calls read and write. Test results indicate that elapsed times decrease by about 30 per cent. This is due mainly to the decrease of the system time, caused by the read and write calls. Objects of the class Node are not created by the C++ constructor function, but are mapped directly on pages in the index file. A special object, the PageManager, administrates which pages in the index file are used and which are free. The PageManager also extends the index file when there are no more free pages available.

The physical page size on a Sun 3/60 is 8,192 bytes. As this is quite large, we used smaller 'logical pages' with sizes of 256 or 1,024 bytes. A node of the index structure does not necessarily occupy an entire logical page. There are usually a few empty bytes left. As the nodes are stored consecutively in the index file, some nodes may cross a physical page boundary. This problem can be solved by adding a field unused bytes to the data-part of the class Node. This solution results in 5–10 per cent shorter query times.

There are several different methods for returning the result of a query to the application program. We implemented two of them: results are returned in one big set, or in a first–next approach. The latter is more suited in interactive applications, but introduces some extra overhead. The drawback of the 'big set' solution is that a (possibly large) temporary structure has to be created (and deleted afterwards). In order to keep the performance tests of the region search queries as pure as possible, we only counted the number of objects that were found. As a final remark, we implemented the quadratic method (Guttman 1984) for splitting a node of the R-tree and the Sphere-tree.

Table 4.1. Created Index Structures for WDB II (31,905 objects) and Art (100,000 objects)

INDEX TYPE	PAGE SIZE (IN BYTES)	MAXIMUM NUMBER OF ENTRIES (M)	MINIMUM NUMBER OF ENTRIES (m)
R1 R-tree	256	12	6
R2 R-tree	1,024	51	26
R3 R-tree	1,024	51	2
S1 Sphere-tree	256	13	7
S2 Sphere-tree	1,024	61	31
S3 Sphere-tree	1,024	61	2
K1 KD2B-tree	256	15 (M_l), 10 (M_i)	8 (m_l)
K2 KD2B-tree	1,024	63 (M_l), 42 (M_i)	32 (m_l)
K3 Packed KD2B-tree	1,024	63 (M_l), 42 (M_i)	–

Table 4.2.   Index Tree Size for WDB II (31,905 objects) and Art (100,000 objects)

INDEX	WDB II	ART
R1	1,163,520	3,457,280
R2	1,016,832	3,048,448
R3	1,623,040	5,760,000
S1	1,106,176	3,170,560
S2	877,568	2,638,848
S3	1,344,512	4,146,176
K1	1,097,984	2,777,344
K2	877,568	2,417,664
K3	541,696	1,696,768

Table 4.3. Point and Rectangle Performance Test
Results WDB II: Average Query Time (over 2,000
queries) per Qualifying Object in CPU ms for the
Index Structures as defined in Table 4.1.

INDEX	POINT	RECTANGLE			
		0.01	0.1	1.0	10.0
R1	24.27	2.68	0.52	0.16	0.11
R2	37.07	3.52	0.62	0.15	0.08
R3	19.66	**2.28**	**0.42**	**0.12**	**0.08**
S1	31.40	5.51	1.10	0.30	0.17
S2	52.48	7.19	1.37	0.31	0.14
S3	41.20	5.76	1.10	0.27	0.14
K1	18.31	4.50	1.09	0.38	0.23
K2	21.86	4.12	0.92	0.27	0.15
K3	**15.60**	3.73	0.83	0.26	0.15

Table 4.4. Convex Polygon and Sphere Performance Test Results WDB II:
Average Query Time (over 2,000 queries) per Qualifying Object in CPU ms
for the Index Structures as defined in Table 4.1.

INDEX	CONVEX POLYGON				SPHERE			
	0.01	0.1	1.0	10.0	0.01	0.1	1.0	10.0
R1	12.14	3.03	1.30	0.96	2.99	0.70	0.29	0.21
R2	25.39	5.51	1.70	0.95	5.03	1.02	0.31	0.18
R3	13.19	3.23	1.27	0.88	**2.72**	**0.61**	0.24	0.17
S1	18.66	3.83	1.15	0.72	4.54	0.86	0.22	0.12
S2	30.66	5.97	1.44	0.70	7.55	1.29	0.25	0.10
S3	23.76	4.72	1.25	**0.67**	4.96	0.90	0.19	**0.09**
K1	**7.47**	**2.32**	**1.13**	0.79	3.83	0.98	0.45	0.33
K2	9.84	2.83	1.21	0.76	3.92	0.94	0.39	0.27
K3	10.73	2.97	1.30	0.79	2.93	0.62	**0.18**	0.09

Table 4.5. Point and Reactangle Performance Test Results Art: Average Query Time (over 2,000 queries) per Qualifying Object in CPU ms for the Index Structures as defined in Table 4.1.

INDEX	POINT	RECTANGLE			
		0.01	0.1	1.0	10.0
R1	188.41	5.09	0.91	0.26	0.14
R2	369.02	9.39	1.44	0.29	0.11
R3	140.04	3.99	0.72	0.20	**0.10**
S1	3907.78	103.92	14.11	2.08	0.46
S2	4650.76	89.07	12.09	1.75	0.37
S3	3526.06	68.72	9.41	1.41	0.32
K1	**31.64**	1.62	0.49	0.28	0.22
K2	34.86	**1.49**	0.39	0.19	0.14
K3	34.52	1.61	**0.39**	**0.17**	0.12

4.4   Performance Test Results

This section presents and interprets the query times of the different index structures applied to WDB II and Art. Other data sets produced similar results. Of the tested page sizes, ranging from 128 to 8,192 bytes, only the results of page sizes 256 bytes (R1, S1, K1) and 1,024 bytes (R2, R3, S2, S3, K2, K3) are given here. In general, these are quite good page sizes. An advantage of the KD2B-tree is that its spatial query times are relatively independent of the page size, owing to the fact that the KD2B-tree has a spatial structure in its internal nodes. In the index structures R1, R2, S1, S2, K1, and K2, the minimum number of entries per node m is chosen to be equal to half the maximum number of entries M: $m = \lceil M/2 \rceil$. In the index structures R3 and S3, the minimum number of entries per node is fixed: $m = 2$. This results in more freedom while inserting or deleting entries and in a usually larger but more efficient structure. To index structure K3, the packed KD2B-tree, the minimum number of entries does not apply. Tables 4.1 and 4.2 summarize the characteristics of the different index structures. Note that the actual tree size of the packed KD2B-tree is the smallest by far.

Tables 4.3–4.6 show the average query time per qualifying (selected) object in CPU ms. Typical numbers of qualifying objects per query for search areas of 0.01, 0.1, 1 and 10 per cent are: 4, 40, 400, and 4,000 for WDB II, and 10, 100, 1,000, and 10,000 for Art. For each query type, the best query time is emphasized by putting it in bold face. For page size 1,024 the same results are also presented in a different form in Figs. 4.9

Table 4.6. Convex Polygon and Sphere Performance Test Results Art: Average Query Time (over 2,000 queries) per Qualifying Object in CPU ms for the Index Structures as defined in Table 4.1.

INDEX	CONVEX POLYGON				SPHERE			
	0.01	0.1	1.0	10.0	0.01	0.1	1.0	10.0
R1	26.93	5.79	1.99	1.17	7.15	1.36	0.44	0.26
R2	77.65	13.60	3.28	1.34	14.98	2.44	0.56	0.24
R3	29.99	6.35	2.06	1.13	6.10	1.20	0.38	0.22
S1	357.57	49.14	7.59	1.84	90.40	11.98	1.72	0.36
S2	388.94	53.43	7.91	1.86	84.36	11.09	1.52	0.29
S3	288.67	40.34	6.38	1.60	64.38	8.53	1.21	0.25
K1	**3.19**	**1.35**	**0.93**	**0.80**	1.23	0.35	0.19	0.14
K2	4.38	1.62	1.00	0.83	**1.16**	**0.28**	**0.12**	**0.09**
K3	5.12	1.85	1.13	0.93	1.28	0.30	0.13	0.09

and 4.10. It is not the average query time per qualifying object that is shown, but the total time of the 2,000 queries.

In the case of WDB II, the R-tree is the best index structure for rectangle queries and the KD2B-tree for the other query types. However, the differences are not very large. In the case of Art, the KD2B-tree is by far the best index structure for (nearly) all query types. In general, the query times for the Sphere-tree are slightly disappointing compared with the R-tree. In the case of WDB II, this could be explained partially by the fact that the total area covered by the MBSs is 2.04 times as large as the total area covered by MBRs. This is due to the 'polyline map' nature of WDB II.

A simple statistical model will clarify this. A polyline is approximated by a straight line segment with length l. Assuming that the angle α which the line segment makes with the positive x-axis is distributed in a uniform manner, i.e. $f(\alpha) = 2/\pi$ for $0 \le \alpha \le \pi/2$, the area of the MBS is independent of α: $\pi l^2/4$. The area of the MBR is $l \cos \alpha \cdot l \sin \alpha$. The expected area of the MBR is 2.46 times smaller than the area of the MBS:

$$A_{MBR} = \int_0^{\pi/2} f(\alpha)(l \cos \alpha \cdot l \sin \alpha)d\alpha = \frac{l^2}{\pi} = \frac{4}{\pi^2} A_{MBS} \approx A_{MBS}/2.46$$

However, in case of 'polygon' maps better results are obtained. The statistical model for general polygons is very complicated. In order to provide some insight we will analyse the MBRs and the MBSs of a square and a rectangle both rotated by angle α. The results for a square are obtained in a similar manner as in the line segment case: $A_{MBS} = \pi l^2/2$ (independent of α) and $A_{MBR} = l^2 (\sin\alpha + \cos\alpha)^2$ with l the length of an edge of the square and α defined and distributed as above. The expected area of the MBR is $(1 + 2/\pi)l^2 = A_{MBS}(2\pi + 4)/\pi^2 \approx A_{MBS}/0.96$, which is larger than the area of the MBS. A slightly more complicated analysis shows that, as long as the length of a side is at least 0.69 times the length of a 'neighbour' side (i.e. the rectangle is not 'too flat'), the MBS is smaller than the expected MBR. Another important factor is the cost of the 'overlap' function. Checking for rectangle–rectangle overlap is the fastest, while the rectangle–sphere and sphere–sphere overlap checks are two to three times slower. The overlap functions for convex polygons are the slowest, twenty to thirty times slower than the rectangle–rectangle overlap.

The test results of Art, where $A_{MBS} \approx A_{MBR}$, indicate that there must be still another reason for the disappointing results of the Sphere-tree. Probably the most important one is that it has less freedom in fitting the MBSs. In the 2D case the MBRs are determined by specifying four parameters, but the MBSs use only three parameters. Together with the fact that the Sphere-tree does not use split lines, this results in a less efficient search structure. An alternative to using MBSs is using minimal bounding ellipsoids (MBEs). These 2D MBEs are also determined by four parameters.

4.5   Discussion

All the structures presented can be used for k-dimensional data, though only 2D test results are shown. The search efficiency of the KD2B-tree met our expectations, that of the Sphere-tree did not. Additional advantages of the KD2B-tree are that the spatial query times are relatively independent of the page size and that the index structure is very small (especially in the case of the packed KD2B-tree). It is possible to design a structure, the dynamic KD2B-tree, that combines the dynamic capabilities of the Sphere-tree with the good properties of the KD2B-tree. For some applications, an axes-parallel version of the KD2B-tree might be useful, because the rectangular query times are expected to be even shorter. This was actually done by Ooi *et al.* (1990, 1987, 1989) in their spatial KD-tree. Another major difference with the KD2B-tree is that the spatial KD-tree is not height-balanced, but its internal nodes are at least filled with $m_i = \lceil M_i/2 \rceil$ entries.

Another aspect of our research is the inclusion of detail levels in the index structure. This will be described in the next two chapters. Working with large map data sets showed the need for this kind of structure, because when the whole region is selected, too much detail is displayed and this takes too much time. Only the most important objects should be selected.

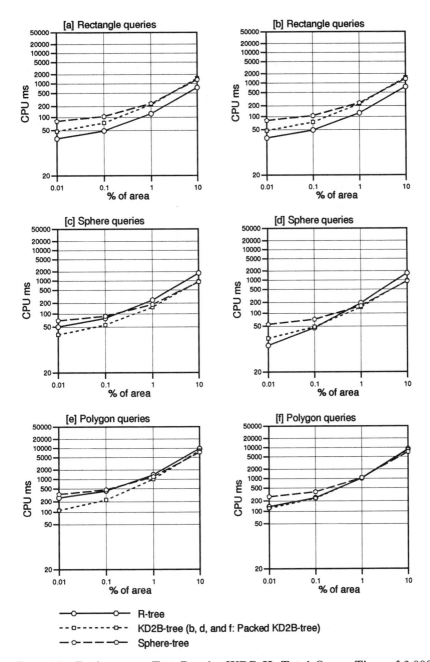

FIG. 4.9. Performance Test Results WDB II: Total Query Time of 2,000 Queries in CPU ms (a, c, and e: $m = \lceil M/2 \rceil$; b, d, and f: $m = 2$)

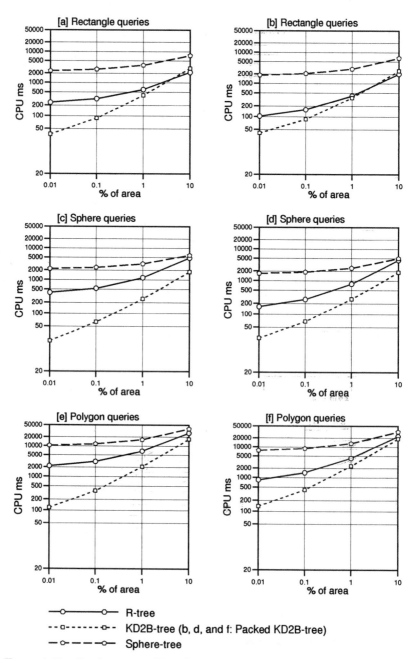

FIG. 4.10. Performance Test Results Art: Total Query Time of 2,000 Queries in CPU ms (a, c, and e: $m = \lceil M/2 \rceil$; b, d, and f: $m = 2$)

5 The Binary Line Generalization Tree

This chapter presents a new data structure with detail levels, which is called the Binary Line Generalization (BLG) tree. Section 5.1 explains why line generalization is required and why known structures are not sufficient. The BLG-tree and an error analysis of the BLG-tree are given in Sections 5.2 and 5.3, respectively. A feature of the BLG-tree, useful in combination with a topological structure, is the so-called join of BLG-trees as described in Section 5.4. The last section describes the limitations of the BLG-tree together with some alternatives.

5.1 Line Generalization

When a small-scale map (large region) has to be displayed, only global and important polylines (or polygons) are selected out of a large-scale geographic data set. However, without specific measures, these polylines are still drawn with too much detail, because all points that define the polyline are used. This detail will be lost on this small-scale because of the limited resolution of the display. Also, the drawing will take an unnecessarily long period of time. It is better to use fewer points. This can be achieved by the k-th point algorithm, which only uses every k-th point of the original polyline for drawing. The first and the last points of a polyline are always used. This is to ensure that the polylines remain connected to each other in the nodes of a topological data structure. This algorithm can be executed when required because it is very simple. The k can be adjusted to suit the specified scale. However, the method has some disadvantages:

- The shape of the polyline is not optimally represented. Some of the line characteristics may be lost if the original polylines contain very sharp bends or long straight line segments.
- If two neighbouring administrative units are filled, for example in case of a choropleth, and the k-th point algorithm is applied on the contour, then these polygons may not fit. The contour contains the renumbered points of several polylines.

Therefore, a better line generalization algorithm has to be used, for instance the Douglas–Peucker algorithm (Douglas and Peucker 1973). Duda and Hart (1973) describe an algorithm similar to the Douglas–Peucker algorithm and call it the 'iterative end-point fit' method. A slightly earlier publication is given by Ramer (1972). These types of algorithms are time-consuming, so it is wise to compute the generalization information for each

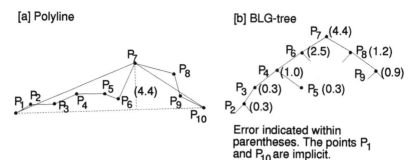

[a] Polyline

[b] BLG-tree

Error indicated within parentheses. The points P_1 and P_{10} are implicit.

FIG. 5.1. A Polyline and its BLG-tree

polyline in a pre-processing step. The result is stored in, for instance, a Multi-Scale Line Tree (Jones and Abraham 1986, 1987). The disadvantages of the Multi-Scale Line Tree have already been discussed in Section 2.3: it introduces a discrete number of detail levels, and the number of children per node is not fixed.

Strip trees (Ballard 1981) and Arc trees (Günther 1988) are binary trees that represent curves (in a 2D plane) in a hierarchical manner with increasing accuracy in the lower levels of the tree. These data structures are designed for arbitrary curves and not for simple polylines. Therefore, I shall introduce a new data structure that combines the good properties of the structures mentioned here. This is the Binary Line Generalization (BLG) tree.

5.2   The BLG-tree

The BLG-tree stores the result of the Douglas–Peucker algorithm in a binary tree. The original polyline consists of the points p_1 to p_n The most coarse approximation of this polyline is the line segment $[p_1, p_n]$. The error for this approximation is determined by the point of the original polyline that has the largest distance to this line segment. Assume that this is point p_k with distance d (see Fig. 5.1 [a]). p_k and d are stored in the root of the BLG-tree, which represents the line segment $[p_1, p_n]$. The next approximation is formed by the two line segments $[p_1, p_k]$ and $[p_k, p_n]$. The root of the BLG-tree contains two pointers to the nodes that correspond with these line segments. In the 'normal' situation, this is a more accurate representation.

The line segments $[p_1, p_k]$ and $[p_k, p_n]$ can be treated in the same manner, with respect to their part of the original polyline, as the line segment $[p_1, p_n]$ to the whole polyline. Again, the error of the approximation by a line segment can be determined by the point with the largest distance. And again, this point and distance are stored in a node of the tree that represents a line segment. This process is repeated until all points are stored in the BLG-tree. Therefore, the BLG-tree incorporates an exact representation of the original polyline. The BLG-tree is a static structure with respect to

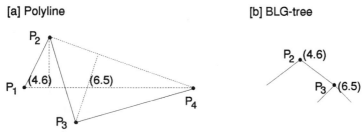

[a] Polyline
[b] BLG-tree

FIG. 5.2. Increasing Error in BLG-tree

inserting, deleting, and changing points that define the original polyline. The BLG-tree of the polyline of Fig. 5.1 [a] is shown in Fig. 5.1 [b]. In most cases, the distance values stored in the nodes will become smaller when descending the tree. Unfortunately, this is not always the case, as shown in Fig. 5.2. The error values on the path from the root to a leaf are not monotonically decreasing.

The BLG-tree is used during the display of a polyline at a certain scale. One can determine the maximum error that is allowed at this scale. During traversal of the tree, one does not have to go any deeper in the tree once the required accuracy is met. The BLG-tree can also be used for other purposes, for example:

- estimating the area of a region enclosed by a number of polylines;
- estimating the intersection(s) of two polylines. This is a useful operation during the calculation of a map overlay (polygon overlay).

5.3 Error Analysis

As indicated above, the BLG-tree is used for the efficient manipulation of digital map data at multiple scales. At smaller scales this will introduce some inaccuracies. During the error analysis, it is assumed that the most detailed data, stored in the BLG-tree form, is the exact representation of the mapped phenomenon. Three topics will be considered in this context: position, circumference, and area of a region enclosed by a number of polylines. The BLG-tree is traversed until a node is reached with an error below a certain threshold, say E. This is always possible because the error value of a leaf node is 0. This traversal results in an approximation of the region by polygon P, defined by points p_1 to p_n, with coordinates (x_i, y_i) for i from 1 to n and $p_{n+1} = p_1$. The estimated circumference C and area A based on polygon P are (van Oosterom 1988a, 1988b):

$$C = \sum_{i=1}^{n} d(p_i, p_{i+1})$$

$$A = \frac{1}{2} \sum_{i=1}^{n} (x_i y_{i+1} - y_i x_{i+1})$$

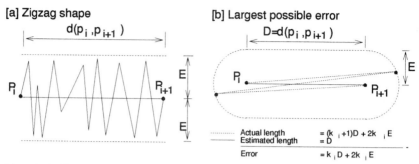

FIG. 5.3. Error in Circumference Estimation

with d the distance between two points. By definition, the position of the region is always within a distance E. If E (in world coordinates) is chosen to correspond to the width of a pixel on the screen, then this will result in a good display function. One can trade efficiency for accuracy and choose E to correspond to the width of three pixels. The accuracy is less, but the display is faster because fewer points are used.

The estimate of the circumference will always be too small. This is because, for every point skipped, two line segments are replaced by one line segment, which has a smaller length than the sum of the replaced line segments. It is hard to say something about the accuracy of the estimation of the circumference C because the boundary can be zigzag-shaped (Fig. 5.3 [a]). If the total number of points in the exact representation of the region is known, the worst-case error can be calculated. For the approximation line segment $[p_i, p_{i+1}]$, with k_i points in between in the exact representation, the largest possible error is (see Fig. 5.3 [b]):

$$k_i d(p_i, p_{i+1}) + 2k_i E$$

When summed over all the approximation line segments, the result is at least bounded by $KC + 2KE$, with C the estimate of the circumference and $K = \sum_{i=1}^{n} k_i$. This is an inaccurate estimate because the error can be more than K times as large as the estimate itself. The BLG-tree is therefore not suited to make estimates of the circumference. It should be noted that in 'normal' cases the estimate may be quite fair in comparison with the worst-case.

The estimate of the area A can be calculated with reasonable accuracy. The largest error per approximation line segment $[p_i, p_{i+1}]$ is (see Fig. 5.4 [a]):

$$2E d(p_i, p_{i+1}) + \pi E^2$$

Summed over all line segments, this gives $2EC + n\pi E^2$, with n the number of points in the approximation polygon P. A better calculation of the worst-case error can be made if one realises that it is impossible for all approximation line segments to have their worst-case situation simultaneously. For a convex polygon (see Fig. 5.4 [b]), the largest possible error is

[a] Error per line segment   [b] Total error

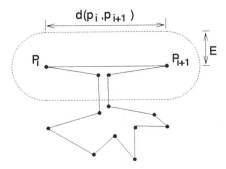 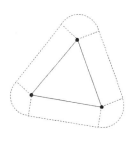

FIG. 5.4. Error in Area Estimation

$EC + \pi E^2$, which is about twice as good as the previous one. The same, though less obvious, is valid for a concave polygon.

5.4   The Join of BLG-trees

On small-scale maps it is possible that a region is represented only by the start- and end-points of the corresponding polylines. As the polylines are defined on the most detailed level (in a multi-scale database), this representation can be too detailed. The BLG-tree should be built for the whole polygon of the region instead of a BLG-tree for each part of the boundary. However, if the results are stored on a region basis, then this will introduce redundant data storage. On the other hand, if the BLG-tree is not stored, then a time-consuming line generalization algorithm has to be executed over and over again.

The solution for this dilemma is the dynamic joining of BLG-trees, which are stored in the polylines. Fig. 5.5 shows how two BLG-trees which belong to consecutive parts of the boundary are joined. The error in the top node of the resulting BLG-tree is not calculated exactly, but is estimated on the basis of the errors in the two subtrees:

$$e_t = d(p_2, [p_1, p_3]) + \max(e_1, e_2)$$

where $d(p_2, [p_1, p_3])$ is the distance from point p_2 to the line segment $[p_1, p_3]$ and e_t, e_1, and e_2 are the error values in the top nodes of the joined and the two BLG-subtrees. The estimate of the error e_t is too large, but this is not a serious problem: the algorithm will never miss any point. The advantage is that it can be computed very rapidly.

The join, in pairs, of BLG-trees is repeated until the whole region is represented by one BLG-tree. Which pair is selected for a join depends on the sum of the error values in the top nodes of the two BLG-trees. The pair with the lowest sum is joined. The building of a BLG-tree for a region is a simple process; only a few joins are needed, one less than the number

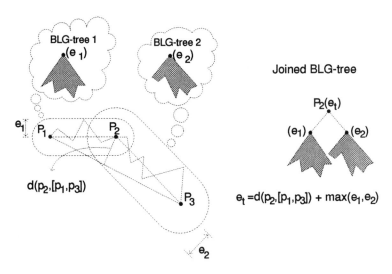

FIG. 5.5. Joining of BLG-trees

of polylines by which the region is represented. Therefore, the BLG-tree for the region can be computed each time it is needed. Using this tree, an appropriate representation for the region can be made on every scale.

5.5   Discussion

The BLG-tree is especially suited for polylines defined by a large number of points. For polylines defined by a small number of points, line generalization computed when required may be more efficient. For polylines that are somewhere in between, another alternative might be interesting. Assign a value to each point to decide whether the point is used when displaying the polyline at a certain scale. This simple linear structure is probably fast enough for the medium-sized polyline. Finally, it must be noted that a BLG-tree is only a part of a reactive data structure. The next chapter describes a more complete reactive data structure, into which the BLG-tree can be integrated.

6 The Reactive-Tree

This chapter presents the first fully dynamic and reactive data structure for efficiently storing and retrieving geometric objects at multiple detail levels. Geometric selections can be interleaved by insertions of new objects and by deletions of existing objects. Detail levels are closely related to cartographic map generalization techniques. The proposed data structure supports the following generalization techniques: simplification, aggregation, symbolization, and selection (see Section 2.3 and Ormeling and Kraak 1987, Robinson *et al.* 1984, Shea and McMaster 1989). Fig. 6.1 illustrates the generalization process by showing the same part of a 1:25,000 map and of an enlarged 1:50,000 map.

The core of the reactive data structure is the Reactive-tree, a geometric index structure that also takes care of the selection-part of the generalization. Other aspects of the generalization process are supported by introducing associated structures, e.g. the BLG-tree (van Oosterom and van den Bos 1989*a*, 1989*b*), for simplification; see Chapter 5. The proposed structure forms an important step in the direction of the development of a seamless, scaleless geographic database (Chrisman 1990*a*, Guptill 1989*b*).

Two geometric data structures that provide some limited facilities for multiple detail levels are the Field-tree (Frank 1983, Frank and Barrera 1989) and the Reactive BSP-tree (van Oosterom 1989, 1990*a*); see Section 2.1 and Chapter 3 respectively. However, these are not fully dynamic. The guideline that important objects must be stored in the higher levels of the tree (see Section 3.7) is the starting point for the design of the Reactive-tree. This guideline was derived during the development of the Reactive BSP-tree. The properties of the Reactive-tree are described in Section 6.1, together with a straightforward Search algorithm. Insert and Delete algorithms are given in the subsequent section. Support for the generalization techniques simplification, aggregation, and symbolization is discussed in Section 6.3. In Section 6.4 the Alternative Reactive-tree is presented, not based on the guideline stated above. The chapter concludes with an evaluation of the presented structures.

6.1 The Properties of the Reactive-tree

In the following subsection, it is argued that importance values associated with objects are required. The two subsequent subsections give an introduction to the Reactive-tree and a formal description of its properties. The last subsection describes a geometric Search algorithm, which takes the required importance level into account.

FIG.   6.1.   The Map Generalization Process (© Topographic Service, Emmen)

6.1.1   Importance Values

Generalization is, stated simply, the process of creating small-scale (coarse) maps out of detailed large-scale maps. One aspect of this process is the removal of unimportant and often, but not necessarily, small objects. This can be repeated a number of times, each time resulting in a smaller-scale map with fewer objects in a fixed region. Each object is assigned a logical importance value, a natural number, in agreement with the smallest scale on which it is still present. Less important objects get low values, more important objects get high values. The use of importance values for the selection of objects was first published by Frank (1981).

Which objects are important depends on the application. In many applications a natural hierarchy is already present. In the case of, for example, a road map, these are: dual carriageways, major four-lane roads, two-lane roads, undivided roads, and unpaved roads. Another example can be found in WDB II (Gorny and Carter 1987), where lakes, rivers, and canals are classified into several groups of importance. Typically, the number of levels is between five and ten, depending on the size and type of the geographic data set. In a reasonable distribution the number of objects having a certain importance is one or two orders of magnitude larger than the number of objects at the next higher importance level, a so-called hierarchical distribution.

6.1.2   Introduction to the Reactive-tree

Several existing geometric data structures are suited to be adapted for the inclusion of objects with different importance values, for example the R-tree (Guttman 1984), the Sphere-tree, and the dynamic KD2B-tree (van Oosterom and Claassen 1990). In this chapter the Reactive-tree is based on the R-tree, because the R-tree is the best-known structure. However, if orientation-insensitivity is important, then one of the other structures mentioned should be used.

The Reactive-tree is a multi-way tree in which, normally, each node contains a number of entries. There are two types of entries: object-entries and tree-entries. The internal nodes may contain both, in contrast to the R-tree. The leaf nodes of the Reactive-tree contain only object-entries. An object-entry has the form

$$(\text{MBR, imp-value, object-id})$$

where MBR is the minimal bounding rectangle, imp-value is a natural number that indicates the importance, and object-id contains a reference to the object. A 2D MBR takes four floating-point numbers, e.g. 16 bytes. The imp-value takes 1 byte, which is enough for 256 importance levels. The object-id takes 4 bytes with the possibility to address over 4,000,000,000 objects. So, in a realistic 2D implementation of the Reactive-tree, the size of an object-entry is 21 bytes. A tree-entry has the form

$$(\text{MBR, imp-value, child-pointer})$$

where child-pointer contains a reference to a subtree. In this case MBR is the minimal bounding rectangle of the whole subtree and imp-value is the importance of the child-node incremented by 1. The importance of a node is defined as the importance of its entries. The size of a tree-entry is the same as that of an object-entry. When one bit in the object-id/child-pointer is used to discriminate between the two entry types, there is no physical difference between them in the implementation and it is still possible to address over 2,000,000,000 objects, which is enough for practical applications. Each node of the Reactive-tree corresponds to one disk page. As in the R-tree, M indicates the maximum number of entries that will fit in one node, and $m \le \lceil M/2 \rceil$ is the minimum number of entries. Assuming that the page size is 1,024, M is 48 in a realistic implementation.

6.1.3 Defining Properties

In this subsection the defining properties of the Reactive-tree are presented. The fact that the empty tree satisfies these properties, and that the Insert and Delete algorithms given in Section 6.2 do not destroy them, guarantees that a Reactive-tree always exists. The Reactive-tree satisfies the following properties:

1. For each object-entry (MBR, imp-value, object-id), MBR is the smallest axes-parallel rectangle that geometrically contains the represented object of importance imp-value.
2. For each tree-entry (MBR, imp-value, child-pointer), MBR is the smallest axes-parallel rectangle that geometrically contains all rectangles in the child node and imp-value is the importance of the child-node incremented by 1.

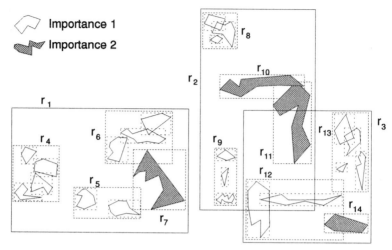

FIG. 6.2. The Scene and the Rectangles of the Reactive-tree

3. All the entries contained in nodes on the same level are of equal importance, and more important entries are stored at higher levels.
4. Every node contains between m and M object-entries and/or tree-entries, unless it has no brothers (a pseudo-root).
5. The root contains at least two entries, unless it is a leaf.

It is not difficult to see that the least important object-entries of the whole data set are always contained in leaf nodes on the same level. In contrast to the R-tree, leaf nodes may also occur at higher levels, owing to the more complicated balancing criteria that are required by the multiple importance levels; see properties 3, 4, and 5. Further, these properties imply that, in an internal node containing both object-entries and tree-entries, the importance of the tree-entries is the same as the importance of the object-entries. Fig. 6.2 shows a scene with objects of two importance levels: objects of importance 1 are drawn in white, and objects of importance 2 are drawn in grey. The figure also shows the corresponding rectangles as used in the Reactive-tree. The object-entries in the Reactive-tree are marked with a circle in Fig. 6.3 for this example. The importance of the root node is 3, and the importance of the leaf nodes is 1.

6.1.4　Geometric Searching with Detail Levels

The further one zooms in, the more tree levels must be addressed. Roughly stated, during map generation based on a selection from the Reactive-tree, one should try to choose the required importance value such that a constant number of objects will be selected, satisfying the rule of thumb on constant pictorial information density. This means that if the required region is large only the more important objects should be selected, and if the required region is small the less important objects should be selected also. The recursive Search algorithm to report all object-entries that have at least

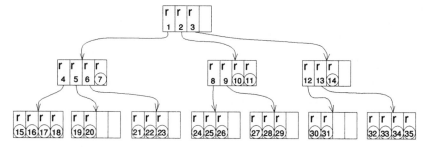

FIG. 6.3. The Reactive-tree

importance imp and whose MBRs overlap search region S is invoked with
the root of the Reactive-tree as current node:

1. If the importance of the current node N is less than imp, then there
   are no qualifying records in this node or in one of its subtrees.
2. If the importance of the current node N is greater than or equal to
   imp, then report all object-entries in this node that overlap S.
3. If the importance of the current node N is greater than imp, then
   also invoke the Search algorithm for the subtrees that correspond to
   tree-entries that overlap S.

6.2 Insert and Delete Entry Algorithms

The Search algorithm is the easy part of the implementation of the Reactive-
tree. The hard part is presented by Insert and Delete algorithms that
maintain the properties of the Reactive-tree. In the implementation presented here, there is exactly one level in the Reactive-tree for each importance value, in the range from min_imp to max_imp, where min_imp and
max_imp correspond to the least and the most important object, respectively. If necessary, there may be one or more tree levels on top of this,
which correspond to importance levels max_imp + 1 and higher. Then the
top level nodes contain tree-entries only. Assume that tree_imp ≥ max_imp
is the importance of the root of the Reactive-tree; then the height of the
tree is tree_imp + 1 − min_imp. The values of min_imp and tree_imp are
stored in global variables. In the algorithms described below, the trivial aspects of maintaining the proper values of these variables are often ignored.
Because of the direct relationship between the importance and the level of
a node in the Reactive-tree of this implementation, the imp_value may be
omitted in both the object-entry and the tree-entry.

6.2.1 Insert Entry

The Insert algorithm described below does not deal with the special cases:
empty tree, and insertion of an entry with importance greater than tree_imp.
Solutions for both cases are easy to implement and they set the global variable tree_imp to the proper value. The Insert algorithm to insert a new
entry E of importance E_imp in the Reactive-tree is:

1. Descend the tree to find the node, which will be called N, by recursively choosing the best tree-entry until a node of importance E_imp or a leaf is reached. The best tree-entry is defined as the entry that requires the smallest enlargement of its MBR to cover E. While moving down the tree, adjust the MBRs of the chosen tree-entries on the path from the root to node N.
2. In the special case that node N is a leaf and the importance N_imp is greater than E_imp, a linear path (with length N_imp $-$ E_imp) of nodes is created from node N to the new entry. Each node in this path contains only one entry. This is allowed, because these are all pseudo-roots.
3. Insert the (path to the) new entry E in node N. If overflow occurs, split the node into nodes N and N' and update the parent. In case the parent overflows as well, propagate the node-split upward.
4. If the node-split propagation causes the root to split, increment tree_imp by 1 and create a new root whose children are the two resulting nodes.

The node splitting in step 3 is analogous to the node splitting in the R-tree. A disadvantage of the Reactive-tree is the possible occurrence of pseudo-roots. These may cause excessive memory usage in case of a 'weird' distribution of the number of objects per importance level, e.g. where there are more important objects than unimportant objects.

6.2.2   Delete Entry

An existing object is deleted by applying the Delete algorithm:

1. Find the node N containing the object-entry, using its MBR.
2. Remove the object-entry from node N. If underflow occurs, then the entries of the under-full node have to be saved in a temporary structure and the node N is removed. In case the parent also becomes under-full, repeat this process. It is possible that the node-underflow will continue until the root is reached and in that case tree_imp is decremented.
3. Adjust the MBRs of all tree-entries on the path from the removed object-entry back to the root.
4. If underflow has occurred, reinsert all saved entries on the proper level in the Reactive-tree by using the Insert algorithm.

There are three types of underflow in the Reactive-tree: the root contains 1 tree-entry only, a pseudo-root contains 0 entries, or one of the other nodes contains $m-1$ entries. The temporary structure may contain object-entries and tree-entries of different importance levels.

6.3   Support for Other Generalization Techniques

The Reactive-tree reflects only a part of the map generalization process: selection. A truly reactive data structure also deals with other aspects of the generalization process. In this section three more aspects are discussed: simplification, symbolization, and aggregation. These terms may be confusing in the context of the Reactive-tree, because the tree is usually described 'top-down' (starting with the most important objects) and map generalization is usually described 'bottom-up' (starting at the most detailed level). All three generalization techniques are incorporated in the reactive data structure by considering objects not as a simple list of coordinates, but as more complex structures. In practice, this can be implemented very well by using an object-oriented programming language, such as Procol; (see Chapters 9 and 10) and (Laffra and van Oosterom 1991, van den Bos and Laffra 1989, 1991, van Oosterom and Laffra 1990, van Oosterom and van den Bos 1989*a*, 1989*b*).

Simplification is an appropriate technique for polyline and polygon type of objects. Besides coordinates, the object structure also contains the BLG-tree we proposed; (see Chapter 5) and (van Oosterom and van den Bos 1989*a*, 1989*b*). By traversing the BLG-tree with the proper error value epsilon, a good graphic representation is obtained. The BLG-tree is most useful for polylines and polygons defined by a large number of points. For a small number of points, executing the Douglas–Peucker line generalization (1973) algorithm when required may be more efficient.

Symbolization changes the basic representation of a geographic entity; for example, a polygon is replaced by a polyline or point on a smaller-scale map. Besides the coordinates of the polygon, the object structure contains a second representation in the form of a polyline or point. Associated with each representation is a scale range which indicates where the representation is valid. An example of the application of the symbolization technique is a city which is depicted on a small-scale map as a dot and on a large-scale map as a polygon.

The last generalization technique included in the reactive data structure is aggregation, that is the combination of several small objects into one large object. From the 'top-down hierarchical tree' point of view, a large object is composed of several small objects; see Fig. 6.4. The geometric description of the large object and the geometric descriptions of the small objects are all stored, because there is no simple relationship between them. The large object is some kind of 'hull' around the small objects; see Fig. 6.4. Usually, a bounding box around the small objects is a sufficient 'geometric search structure', because the number of small objects is limited. However, if the number of small objects combined in one large object is quite large, then an R-subtree may be used.

Aggregation is used, for example, in the map of administrative units in the Netherlands; see Subsection 1.5.2. Several municipalities are grouped into one larger economic geographic region (EGR), EGRs are grouped into a nodal region, nodal regions are grouped into a province, and so on. Another approach to this case is to consider the boundaries, instead of the regions, as starting point of the design. In that case selection is the appropriate generalization technique and the Reactive-tree can be used without additional structures. Note that a reactive data structure does not provide any explicit topological information. Therefore, the Reactive-tree may be used on top of a topological data structure; see Section 2.2.

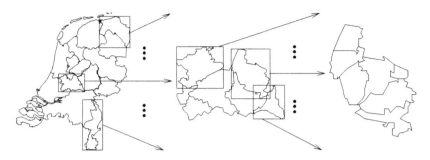

FIG. 6.4. A Large Object Is Composed of Several Small Objects

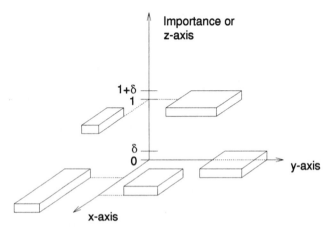

FIG. 6.5. The 3D MBRs of the Alternative Reactive-tree

6.4   An Alternative Reactive-tree

This section presents a reactive data structure that is not based on the guideline that important objects must be stored in the higher levels of the tree. The advantage of the Alternative Reactive-tree over the Reactive-tree is that it does not assume a hierarchical distribution of the number of objects over the importance levels.

The 2D Alternative Reactive-tree is based on a 3D R-tree. The 3D MBR of a 2D object with importance imp is defined by its 2D MBR, and its extents in the third dimension are from imp and to imp $+\delta$, where δ is a positive real number, so an object corresponds to a block with non-zero contents (except for point objects). Fig. 6.5 depicts the 3D MBRs of a number of 2D objects at two different importance levels. When the parameter δ is set to a very low value, e.g. 0.01, the Alternative Reactive-tree tries to group the objects that belong to the same importance level. This can be explained by the fact that there is a heavy penalty on the inclusion of an object with another importance value, as the volume of the 3D MBR will increase by at least a factor of $(1+\delta)/\delta$. The larger δ becomes, the less is the penalty, and the more likely it is that objects of different importance are grouped, and the Alternative Reactive-tree behaves more like a normal 2D R-tree. In any case, all objects, important and unimportant, are stored in leaf nodes on the same level.

The Alternative Reactive-tree can be generalized to support objects with general labels instead of the hierarchical importance values. This enables queries such as 'Select all capital cities in region R'. The label capital is associated with some of the geographic objects, by inserting these entries into the tree. A geographic object may be associated with more labels by inserting more entries for the same object. In the implementation, a label corresponds to a numeric value. By choosing certain values for these labels and for δ, possible coherence between labels may be exploited. This is what is actually done in the 2D Alternative Reactive-tree for hierarchically distributed data.

6.5 Discussion

This chapter describes the first fully dynamic and reactive data structure. It is presented as a 2D structure, but 3D and higher-dimensional variants are possible. Note that this has nothing to do with the use of a 3D R-tree for the 2D Alternative Reactive-tree. The Reactive-tree and the Alternative Reactive-tree have been implemented in the same environment as the KD2B-tree, Sphere-tree, and R-tree; see Chapter 4. In analogy to the R-tree, it is also possible to create a Reactive-tree based on the Sphere-tree. An Alternative Reactive-tree based on the Sphere-tree needs no δ, because the 3D MBS (sphere) normally has non-zero contents. However, this structure is useful only when the importance values are chosen far apart relative to the size of the geometric objects. Therefore, the positive real number will be a more suitable data type for this importance value.

Two large data sets have been used to test the reactive structures: WDB II (Gorny and Carter 1987) and the map of administrative units in the Netherlands. Both tests showed the advantage of the selection based on importance level and geometric position. Displaying the whole map area at interactive speed was possible, in contrast to the situation where the normal R-tree was used, which also showed a lot of annoying details. The additional structures for the support of simplification, symbolization, and aggregation are currently being implemented. Future performance tests depend on the availability of digital maps with generalization information.

Two other generalization techniques were not discussed: exaggeration and displacement; see Section 2.3. Exaggeration seems easy to include, because it is a simple enlargement of an aspect of the graphic representation of one object, e.g. the line width. However, the enlargement of linear features may cause other features to be covered and they must therefore be displaced. Exaggeration and displacement are difficult to handle, because multiple objects have to be considered. An *ad* hoc solution is to associate an explicit set of tuples (displacement, map-scale-range) with each object that has to be displaced and a set of tuples (enlargement, map-scale-range) with each object that has to be enlarged. Further research is required in order to develop more elegant solutions.

Very recently, another reactive data structure has been proposed by Becker and Widmayer (1991). The Priority Rectangle File (PR-file, based on the R-file (Hutflesz *et al.* 1990)), forms the backbone of their structure. A significant common characteristic of the PR-file and the Reactive-tree is that, in general, both store more important objects in higher levels. A few differences of the PR-file, compared with the Reactive-tree, are that objects of equal importance (priority) are not necessarily on the same level, and object-entries and tree-entries cannot be stored in the same node. Another recent development in the storage structures for map data that are to be used at different resolutions is presented by Persson and Jungert (1991). Their method is based on structuring run-length encoded (RLC) raster data.

Finally, other Reactive-trees should be considered which are able to deal efficiently with a non-hierarchical distribution of the number of objects over the importance levels, while sticking to the guideline that important objects are to be stored in the higher levels of the tree. This might be realised by changing the properties in such a manner that one tree level is allowed to contain multiple importance levels, but it is not (yet) clear how the Insert and Delete algorithms should be modified. This also is subject to further research.

Part II

Persistence and Object-Oriented Modelling

In the introductory Chapter 1, six requirements for a geographic data model were formulated. Part I of this book concentrated on the requirements $r2$ (spatial properties) and $r3$ (detail levels) for the geographic data model. This resulted in the introduction of reactive data structures. Requirement $r6$ (dynamics) also received considerable attention, as several dynamic data structures were presented.

In this part, solutions for the other requirements will be described. The requirement $r4$ (layers) is not very hard to satisfy and does not pose any problem, because the layers in a geographic data set can be treated quite independently. The integrated storage of geometric, topological, and thematic data in one system ($r1$) is much harder. Therefore, an extension to the relational data model is suggested in Chapter 7, which includes geometric data types, operators, and index structures. The implemetation of this extension is described in the subsequent chapter. The database solution provides long-term storage ($r5$), but there is a large gap between the data stored in the database and the data stored in data structures of a running program. Another drawback is that it is hard to represent complex objects, as introduced by for example aggregation (Chapter 6), in the flat tuples of a relational database.

Chapter 9 shows that an object-oriented programming language, such as Procol, is better suited for the modelling of the object types encountered in GISs. Object instances of different size, e.g. polylines, are no problem. The only disadvantage is that a file or DBMS is still required for the long-term storage of the contents of the objects. We solved this problem by introducing persistent objects in Procol, as described in Chapter 10. Index structures can also be incorporated, which enables fast access to the objects. In addition to its more powerful modelling capabilities, this solution has another advantage over the DBMS solution. The gap between the data in the storage system and the data in the running program is minimized. This means that there is no need for the, often complicated, code that loads and saves the data from and to the storage system. The described implementation technique for Persistent Procol not only makes the development more efficient, but also results in very efficient (fast) programs.

7 A Geographic Extension to the Relational Data Model

Many of today's information systems, including Command, Control, and Communication Information (C^3I), are built around a relational database. However, the relational data model was developed primarily for business data processing systems. In this chapter we will use the Geographic Analysis Intelligence System to illustrate the required functionality of a C^3I system. In Subsection 1.5.4, a description of the functions and geographic requirements of the Geographic Analysis Intelligence System was given. Here I shall emphasize the typical geographic properties of this system. The Geographic Analysis Intelligence System is a modified (but still very realistic) version of a C^3I system that is currently under development for the Royal NetherLands Air Force (RNLAF). To solve the problems that arise from the geographic nature of the system, I propose an extension to the relational data model that includes geographic capabilities. I have chosen the relational data model as a starting point, because it is a powerful and well-known model.

Because of the data requirements of the Geographic Analysis Intelligence System, it seems appropriate to use a relational database for storage and retrieval of the data. An explanation of the relational data model is given in Section 7.1. Reasons for using a relational database in C^3I systems are given in Section 7.2. Having discussed the Geographic Analysis Intelligence System and the relational data model, we will look into possible solutions of the geographic requirements of the system, using this relational data model, in Section 7.3. I will then summarize the problems that arise when using current relational databases in this way. Because we want to stay within the relational data model for a number of reasons (see Subsection 7.3.4), possible geographic extensions to this model are discussed in Section 7.4. We will then reconsider the geographic requirements of the Geographic Analysis Intelligence System, using these extensions. Finally, in Section 7.5, results and recommendations are given.

7.1 The Relational Data Model

This section first gives an overview of the basic concepts of the relational data model, which is the basis of Relational DataBase Management Systems (RDBMSs). Next, a description is given how relational databases can be defined and manipulated, using a relational query language.

empl#	name	salary	street	town
775	van Hekken	20,000	Main Street 3	Delft
816	van Oosterom	20,000	Boulevard 12	Woudenberg
940	Woestenburg	25,000	Station Road 6	Delft

FIG. 7.1. Fictive Contents of a Relation

7.1.1   Mathematical Description

As their name implies, relations play a major part in relational databases. First, a formal definition of a relation (Codd 1970):

> Given a collection of sets $D_1, D_2, \ldots, D_n$, R is a relation on these sets if it is a set of ordered n-tuples $< d_1, d_2, \ldots, d_n >$ with $d_i \in D_i$. The sets D_i (for $1 < i < n$) are called the domains of R and n is the degree of R. The elements d_i are the attributes of the relation.

In the relational database the data types of attribute domains are limited to: fixed-length string, integer, and real. A classic example is the following relation:

employee(**empl#**, name, salary, street, town)

In this example, employee number (empl#) and salary are integers and the other attributes are strings of length 30. A relation consists of a number of tuples or records (attribute domains with actual values) that can be described by a two-dimensional table.

Each row of this table contains a tuple. The columns each describe a certain attribute. Within a relation at least one attribute (or combination of attributes) must have a unique value for each tuple. This attribute is called the primary key; i.e., it identifies the tuple and other attributes that are dependent on it. The primary key may not have an empty value. In our example empl# is the primary key. If an attribute (or a combination of attributes) other than those of the primary key are suitable for identification of the tuple, this attribute is called a candidate key. In the relation 'employee' the attribute name is a candidate key. A candidate key that is not a primary key are called an alternate key. A foreign key is an attribute (or combination of attributes) in a relation whose value must match those of the primary key in another relation. The other attributes are referred to as secondary attributes. Fig. 7.1 shows the fictive contents of a relation.

The normal forms of relations are important during the design of a relational database application (Date 1981). They help to remove redundancy, so that it is easier to keep the database consistent if changes are made. In the process of normalization, the concept of (complete) functional dependency and multi-valued dependency plays an important part. Functional dependency is defined as follows: Attribute b of relation R is functionally dependent on attribute a of R, if and only if every value of a determines exactly one value of b. In this definition the attributes a and b may be a combination of 'atomic' attributes (integer, real, string). Complete functional dependency is defined as follows: Attribute b of relation R is complete functionally dependent on attribute a of R, if attribute b is functionally dependent on attribute a and not on any subset of attribute a. Multi-valued dependency is defined as follows: There is a multi-valued dependency from attribute a of relation R to attribute b of R, if a value of a determines a set of values of b. Both a and b will be part of the (candidate) key of R. Using these definitions, the five normal forms are defined as follows (Date 1981):

- NF1: A relation is in first normal form if and only if all attribute domains contain 'atomic' values.
- NF2: A relation is in second normal form if and only if it is in first normal form and everyone of its (secondary) attributes are completely functionally dependent on the candidate key of the relation.
- NF3: A relation is in third normal form if and only if it is in second normal form and everyone of its secondary attributes are not transitively dependent on the candidate key of the relation.
- NF4: A relation is in fourth normal form if and only if it is in third normal form and does not contain more than one multi-valued dependency.
- NF5: A relation is in fifth normal form if and only if it is in fourth normal form and it cannot be decomposed into three or more smaller tables without loss of information.

Note that the relation 'employee' defined above is in fifth normal form.

7.1.2   The Relational Query Language

To manage relational databases, current RDBMSs contain query languages like SQL (Structured Query Language: see Date 1981), which will be used as an example in the rest of this chapter. The main parts of SQL are the Data Definition Language (DDL) and the Data Manipulation Language (DML). The most important property of the DDL is its ability to define relations in the database (called tables). For example, in Fig. 7.2 [a] the relation 'employee' is defined using the DDL part of SQL.

Using the DML, existing tuples can be deleted from the relation (table) and new ones can be added, using the operations DELETE (respectively INSERT). It is also possible to update the attribute values of existing tuples, using the operation UPDATE. This is not allowed for the primary key. Data can be retrieved from the database by searching in a table, using the operation SELECT. The available operators can be categorized into the following classes (Chang and Fu 1980):

- comparison operators: $=, \neq, <, \leq, >$, and $\geq$;
- logical operators: AND, OR, and NEGATE;

[a] SQL definition of relation "employee"

```
CREATE TABLE EMPLOYEE (  EMPL_NR  NUMBER(3)  NOT NULL,
                         NAME     CHAR(30),
                         SALARY   NUMBER(5),
                         STREET   CHAR(30),
                         TOWN     CHAR(30)  );
```

[b] Example of SQL query

```
SELECT NAME
FROM EMPLOYEE
WHERE TOWN = "Delft"
```

[c] Result of the previous query

```
name

van Hekken
Woestenburg
```

FIG. 7.2. Some SQL Examples

- statistical operators: MIN, MAX, COUNT, TOTAL, and AVERAGE;
- set operators: Cartesian product, union, intersection, and difference;
- special database operators: restriction, projection, and join.

These operators are the basic parts of the query language. The queries can be formulated by an application program (through a program interface to the RDBMS, called embedded SQL) or directly by the user. In this way it is possible to put both simple and more complex questions (called queries) to the RDBMS and to specify the format of the required answer. For example, the query 'Select the names of the employees who live in Delft' is formulated in Fig. 7.2 [b], using SQL. The result of this query is given in Fig. 7.2 [c].

A very important operation is the joining of relations. The attribute of one relation is matched with a corresponding attribute in another table. All combinations make up the tuples of the new relation. For instance, if we define the relation project (**proj#, empl#**) in which both attributes are part of the primary key and create a corresponding table with SQL, which contains the data given in Fig. 7.3 [a], we can ask the question 'Select the names of the employees that are working on project with number 20357'. In SQL, this question is stated as in Fig. 7.3 [b]. The two tables are joined by the 'empl#'. The result of the query is given in Fig. 7.3c.

In the case of selection of tuples from a relation, all tuples have to be

[a] Fictive contents of the relation "project".

```
    proj#       empl#

    20461       775
    20357       775
    20357       816
    20461       940
```

[b] SQL query to select all employee names working
    on project 20357

```
SELECT NAME
FROM PROJECT, EMPLOYEE
WHERE PROJECT.EMPL_NR = EMPLOYEE.EMPL_NR
      AND PROJECT.PROJ_NR = 20357
```

[c] Result of the previous query

```
name

van Hekken
van Oosterom
```

FIG. 7.3. Some More SQL Examples

visited and checked for the right attribute value. This will take a lot of time in a realistic large table and, if required often, will be very annoying. In many commercial implementations, including Oracle (Oracle Corporation 1990), Ingres (Sun Microsystems, Inc. 1987), etc., this problem can be solved by putting an index on often-searched attributes of the relation. This is an additional structure (sorted on the attribute(s)) that contains references to the original tuples. The index is, in general, implemented as a B-tree (Bayer and McCreight 1973), which allows searching in $O(\log n)$ time.

In the next section I will suggest reasons for using a RDBMS in C^{3}I systems. In the subsequent section examples of an implementation of the Geographic Analysis Intelligence System, using current a RDBMS, will be presented.

7.2 The C^{3}I Relational Database

Like many of today's C^{3}I systems, the Geographic Analysis Intelligence System will be built around a RDBMS to store the data. Background maps, although they do not change very often, should, if possible, be stored in the same database. There are a number of reasons for using a RDBMS:

- By using a relational database, data redundancy and data inconsistency are avoided, because every single item is stored only once.
- A relational database ensures a flexible growth path in case of changes in operational concepts and changes in the structure of the data that have to be managed.
- A relational database ensures enough flexibility to support simple and efficient interfacing with other C^{3}I systems.

Current RDBMSs are developed primarily for business data processing. In these systems the data that have to be managed are not geographically oriented and there is generally no need for graphic presentation of information. However, as will be clear from the description in Subsection 1.5.4, the Geographic Analysis Intelligence System does contain these geographic data. Referring to the required geographic functionality of the system, I will describe the problems that arise when we try to develop a pure relational solution for it.

7.3   A Pure Relational Solution of the Problem

The first two subsections will develop a pure relational solution of some geographic aspects of the case, both in the area of data definition and in the area of data manipulation. This or a similar approach is used in several systems described in the literature, including GEO-QUEL (Berman and Stonebraker 1977), a geographic information retrieval and display system built on top of Ingres (Sun Microsystems, Inc. 1987). As we shall see, there are some problems with these solutions, which are summarized in the third subsection. The last subsection will provide reasons for not discarding the relational data model, but instead extending current RDBMSs with geographic capabilities.

7.3.1   Definition of Data

This subsection shows that it is possible to store geographic data of the Geographic Analysis Intelligence System in a RDBMS without any modifications or extensions. The first kind of geographic data that we want to store are the (static) background map data (borderlines of countries, rivers, trafficways, etc). The solution presented here is a slightly simplified version of the data structure presented by Van Roessel (1987), who developed this data structure mainly for the interchange of geographic data. He applies a technique described by H. C. Smith (1985) for developing a set of fully normalized relations to the topological data model. This is a well-known data model (Peucker and Chrisman 1975) that deals with nodes, chains, and polygons (see Fig. 7.4), and hence is very suitable for describing the kind of background data that we want to use. The following relations are defined (**bold** items denote a primary key and *italic* items denote a foreign key):

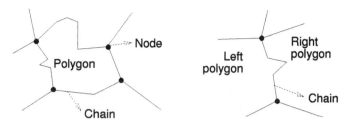

FIG. 7.4. Chains and Nodes

    point(**pntid**, x, y)
    node(**nodeid**, *pntid*)
    chainpoints(**chainid, pntseq,** *pntid*)
    chaintopol(**chainid,** *strnode, endnode, lftpol, rgtpol*)
    polygon(**polid, chnseq,** *chainid*)

The chain concept of the topological data model is captured in two relations: chainpoints and chaintopol. The first of these describes the shape of the chain. As the order of points within a chain is important and tuples within a relation are not ordered, the chainpoints relation must also have an attribute describing this order (the attribute 'pntseq'). The second relation (chaintopol) captures the topological aspects of a chain.

It is surprising that these intuitively clear (and good) relations can be derived by Smith's techniques and are therefore fully normalized. The fact that polygons may have (nested) holes is removed from Van Roessel's example because of clarity. For the same reason, only a single map layer (one geographic theme) is modelled. However, the concept of spatial domain (a limited geographic region of interest) is removed from the model of Van Roessel. In a seamless map (Guptill 1989*b*, van Oosterom 1991) , the domain should, in principle, be infinitely large and not limited, like paper map sheets, to a certain region.

As we have seen, the Geographic Analysis Intelligence System contains a number of dynamic geographic objects, some of them at various levels of detail. Target objects (airbases, buildings, bridges, etc.) and threat objects are described by a single point location. Target areas, planning lines, and airspace management data are described by polygons or polylines. To store these objects, which are (in contrast with, for instance, the background polygons) independent of each other, somewhat simpler relations than the ones given above can be used. For instance, targets and threats that can be described by one point location are defined as:

    target(**tar_id**, pntid, tar_attr_1, ... ,tar_attr_n)
    threat(**thr_id**, pntid, thr_attr_1, ... ,thr_attr_n)

Operations zones that can be described as an area are defined as:

    ops_zone(**oz_name**, oz_attr_1, ... ,oz_attr_n)
    ops_zone_area(**oz_name, oz_loc_seq**, pntid)

This relation contains the locations that form lines that enclose the operations zone (area). If the area is described by k locations, location 1 must be connected (by a line segment) to location 2, location 2 to location 3, ..., and location k to location 1. For this purpose, a location ordering attribute is used ('oz_loc_seq'). Note that pntid is used instead of chainid in the relation ops_zone_area. This is because we are not interested in the topology. When using objects with various detail levels, the attribute pntid (and possibly some other geographic attributes) will normally be included (as a foreign key) in all the relations that describe those detail levels.

7.3.2   Spatial Queries

Apart from the possibility of defining data in current RDBMSs (using the DDL part), it is also possible to query and manipulate those data (using the DML part). All RDBMSs use some query language to retrieve and manipulate data of the database. For this purpose, the query language contains relational operators. Because, as we have seen, the main functions of the Geographic Analysis Intelligence System are entry, updating, storage, retrieval, and presentation of (partly geographic) intelligence and airspace management data, it would be desirable to be able to ask questions like 'Select all threat objects within an operations zone' or 'Select the target objects that are closest to a fixed point location' just as easily as questions like 'Select the names of the employees who live in Delft' or 'Select the employee number of the employee with the highest salary'. However, in current RDBMSs, spatial queries become very complex and difficult. For instance, the question 'Select all threat objects within an operations zone' can be stated relatively easily in SQL when the operations zone is a rectangle denoted by the opposite corners (x_1, y_1) and (x_2, y_2) (see Fig. 7.5 [a]). However, if the operations zone is a polygon of an arbitrary number of points, the SQL query will become very complex (if not impossible). Furthermore, it is not possible to ask the question 'Select the target objects that are closest to a fixed point location' in an easy and natural way in SQL, as proposed in Fig. 7.5 [b]. In this query the fixed-point location is denoted by (x_f, y_f).

In these query examples a join of two relations is necessary, which would not be the case if data types like a single point location were available. It would be even more difficult (if not impossible) to write the distance calculation formula using the available relational operators of SQL. The reason for the complexity or even the impossibility of the geographic queries is obvious. The RDBMS has no knowledge of the way in which the geographic data types (points, polylines, and polygons) are implemented and, as a consequence, has no straightforward relational operators, like DISTANCE, available for these data types, as it has for numbers and character strings. The same problems apply to questions like 'Select all threat objects within a certain operations zone'. In SQL no operation is available to test directly whether a point location is located within an area that is described in a manner like the relation 'ops_zone_area'. Some of the problems that we

[a] Example of selection of threats within rectangle

```
SELECT THREAT_ID
FROM THREAT, POINT
WHERE TARGET.PNTID = POINT.PNTID
      AND    x1 < POINT.X < x2
      AND    y1 < POINT.Y < y2
```

[b] Example of minimum distance determination

```
SELECT MIN(DISTANCE(POINT.X, POINT.Y, xf, yf))
FROM TARGET, POINT
WHERE TARGET.PNTID = POINT.PNTID
```

FIG. 7.5. Spatial Queries

have seen from the examples in this subsection will be summarized in the next subsection.

7.3.3   Problems of the Pure Relational Solution

The major spatial requirements for a database can be stated as follows (Roussopoulos *et al.* 1988): 'The database must support domains which consist of non-atomic non-zero space objects.' Besides storing these objects, this also means that the database must support direct spatial search and computation. As will be clear by now, the spatial capabilities required by the presented case are very hard or even impossible to meet with current RDBMSs. There are several reasons for this.

First, current RDBMSs lack geographic data types. Point locations must be implemented using character strings or numbers. Polylines and polygons are represented by multiple rows in a table with a repetition of the id of the object on each row. Furthermore, a sequence number must be added to each row (denoting a point location), because these would otherwise be unordered. This is all redundant data storage which can be avoided by other representations.

As a second and closely related problem, the manipulation of objects is complex: a difficult query, possibly followed by some processing. This is due to the fact, that because current RDBMSs lack geographic data types, there are no geographic operators available and a very complex combination of the available relational operators for character strings and numbers must be chosen (if possible). Sometimes it is even impossible to perform this processing within the database and special application functions must be provided; i.e., the knowledge of the geographic data structure must be built into the applications. An example is the calculation of the area of a polygon.

Finally, efficient search for spatial objects is impossible. Only one-dimensional search can be performed efficiently, that is in $O(\log n)$ time.

Kemper and Wallrath (1987) describe the same problems in the context of CAD/CAM applications. Despite these problems, in the next subsection it will be argued that it is not convenient and not necessary to abandon the relational data model completely.

7.3.4   Sticking to the Relational Data Model

Although the relational DBMSs are designed primarily for business data processing, the developed database concepts, such as data independence, data integration, controlled redundancy, and security, are equally important for databases with geographic capabilities as we have seen earlier. This is one reason for not discarding the relational model. Another reason is that the relational model is conceptually simple and well-known by large groups of users. In every spatial database there also has to be a way to store the thematic (alpha-numeric) data, so why not use the relational model for this purpose? This approach is chosen by several implementators of commercially available GISs. Because of the problems with the spatial data, their database is often split into two parts (not necessarily noticeable by the end-user): a relational part and a special part for the geometric data. These two parts are linked by corresponding object-ids of the attribute and geometric parts: the dual architecture. It will be clear that it is more elegant to combine the two database parts in one system and thus avoid the problem of linking these together.

In the following section I shall propose some extensions to the relational model to make it more suitable for geographic operations. These extensions will occur in the DDL part of the query language, e.g. additional data types and more than one level of detail, as well as in the DML part of the query language. As we will see, because of these extensions, new index structures must be provided too.

7.4   The Geographic Extension

The actual geographic extension of the relational data model consists of three parts. First, new geographic data types (domains) are introduced. Second, operators are defined for these new data types. Third and finally, new indexing methods are required when dealing with geographic information. Note that in this chapter we concentrate on the two-dimensional extension. However, three-dimensional or higher-dimensional extensions are quite similar.

These extensions will be described at a conceptual level; that is, the exact syntax (of for example the query language) will be omitted. Further, the alpha-numeric interaction technique, typing in queries using a keyboard, is not sufficient when dealing with geographic data. Some kind of combination with graphic interaction is needed, but this is also outside the scope of the research described in this book.

7.4.1 Data Types (or Domains)

Nearly all geographic data processing (Nagy and Wagle 1979, van Oosterom 1988a, 1988b) is performed with the vector, the raster, or a combination of these geometric data formats. The vector format has three subtypes: point, polyline, and polygon. The emphasis is on the vector representation, because it allows more flexible manipulations, although the raster representation has also advantages (see Section 1.2). Though their representations might be complex, these new data types must be regarded as atomic values in the data model. So it is possible to define the following relations:

target(**tar_id**, tar_point, tar_attr_1, ..., tar_attr_n)
threat(**thr_id**, thr_point, thr_attr_1, ..., thr_attr_n)
ops_zone(**oz_id**, oz_polygon, oz_attr_1, ..., oz_attr_n)

Note that the relation 'ops_zone_area' is no longer necessary. The relations are at least in the first normal form, because they contain only atomic attributes. If the new data types point, polyline, and polygon were not considered atomic, the data model would be non-normalized. The data types of tar_point, thr_point, and oz_polygon are Point2 and Polygon2, respectively. The '2' behind Point and Polygon indicates that this is a two-dimensional attribute: Polygon2 does not mean that this is a polygon with two points! In the same manner, it is possible to use these data types in a three-dimensional variant. The circle is not included as a basic data type, though it is one of the intelligence objects; see Subsection 1.5.4. The rationale behind this is that the proposed extension must be compact. The introduction of a circle would also require operators for the circle. If a circle is introduced, why not also introduce an ellipse, a spline, and so on? The circle can be approximated by a polygon. A more advanced extension might also include the data types procedure and reference to another relation (query). Both of these data types are useful for the implementation of detail levels. Associated with a polyline or polygon is a line generalization algorithm to reduce the number of points used when working with small-scale maps. Another method of dealing with detail levels is to allow references to other tables, describing the refinement of objects on a larger-scale map (Chang and Kunii 1981).

7.4.2 Spatial Operators

Only the basic operators will be described. More complex operators are not part of the database system, but belong in a specific application, for example network calculations (shortest path, travelling salesman, location of service centre), advanced visualization techniques (prism maps, digital elevation models), or simulations (evacuation of a region by road, flow of liquids through pipes). The polygon overlay takes two sets of polygons and calculates all intersections, which results in a third set of polygons. Though it is a complex operation, it is used in many GIS applications. So, perhaps it should also be in the set of standard operators.

Many spatial operators have been described by various authors (Chang and Fu 1980, Güting 1988*b*, Joseph and Cardenas 1988, Menon and Smith 1989, Raper and Bundock 1990, Roussopoulos *et al.* 1988). I do not claim that my lists of operators are complete, but they should give a good impression of the basic spatial operators. Three fundamental classes of spatial operators are distinguished in addition to the five classes of Subsection 7.1.2. These are: spatial comparison operators, geometric calculations operators, and topological operators. Also, there must be an 'output operator' that directly displays the result of a geographic query on a graphics display.

1.   Geometric calculation operators return a real/integer value or geometric value. Some of the most important operators are: DISTANCE (operands: two points, polylines or polygons), LENGTH (operand: one polyline), PERIMETER (operand: one polygon), AREA (operand: one polygon), CLOSEST (operands: one point, polygon, or polyline and a set of candidate objects), INTERSECTION (operands: two polylines or polygons), UNION (operands: two polylines or polygons).

2.   Topological operators return a geometric value. Some examples: neighbours, next link (in a polyline network), left and right polygons of a polyline, start and end nodes of polylines. We do not need a special set of topological operators, because the topological model can be captured in a very natural manner in the standard relational model; see Subsection 7.3.1.

3.   Spatial comparison operators return a Boolean (TRUE or FALSE). Though the actual calculations are often (partly) the same as in the previous classes of operators, they form a separate group, for example ON, INTERSECTS, INSIDE, OUTSIDE, NORTH_OF, NEIGHBOUR_OF, LARGER_THAN. All comparison operators have two operands, which may be, in most cases, of any geometric type. Note that it is often possible to emulate these operators by combining the normal comparison operators and the other geographic operators.

There are always more spatial operators, which might be useful for a specific application. But it is impossible to include them all. An 'open' database is the solution for this problem. If an operator is not within the set of basic spatial operators, it is implemented (by the user), and after it has been certified it is added to the database system. In this way other users also benefit from the new capabilities. Note that organizational actions have to be taken; e.g., someone must have the responsibility for clear and unique names of operators. Examples of non-basic spatial operators include: calculation of the Voronoi diagram, the convex hull, the smallest enclosing circle, and so on. These are pure geometric problems, but less common than the ones in the set of basic operators. The research in the relatively new area of science called 'Computational Geometry' (Blankenagel and Güting 1988, Edelsbrunner 1987, Kriegel *et al.* 1991, Preparata and Shamos 1985) offers very efficient techniques for implementing the (basic and non-basic) spatial operators. Regardless of the implementation, it is possible to formulate queries with the basic operators (in SQL-like syntax) as in Fig. 7.6.

[a] Reformulation of "minimum distance" query

```
SELECT MIN(DISTANCE(TAR_POINT, point))
FROM TARGET
```

[b] Same query, but formulated more efficiently

```
SELECT CLOSEST(TAR_POINT, point)
FROM TARGET
```

[c] Query using INSIDE operator

```
SELECT THR_ID
FROM    THREAT
WHERE   INSIDE((SELECT OZ_POLYGON
                FROM OPS_ZONE
                WHERE OZ_ID = 12), THR_POINT)
```

FIG. 7.6. Examples of Queries using 'Extended' Operator Set

7.4.3 Indexing Techniques

The B-tree, an indexing technique used in many database implementations, combines several desirable properties. It is a dynamic, height-balanced structure; that is, inserts, deletes, and updates of entries may be interchanged by searches. Because of the balanced nature, searches are efficient (in $O(\log n)$ time). Also, the B-tree has a high memory occupation rate (Bayer and McCreight 1973).

However, the B-tree is suited only for searching based on one-dimensional (number/string) attributes. Multiple indexes on more than one attribute of a relation are possible, but (with current implementations of RDBMSs) only one can be used for solving a query like: 'Select all employees with empl# > 700 and $15,000 <$ salary $< 22,500$' or 'Select all threats in the region $5 < x < 10$ and $12 < y < 20$'. These point queries can be solved efficiently by the KDB-tree (Robinson 1981). The KDB-tree is a KD-tree adapted for secondary storage and it can handle point data from any dimension. The KD-tree cannot handle the other geometric data types: polyline and polygon. In the literature there are several solutions for this problem, such as the R-tree (Faloutsos *et al.* 1987), the Field-tree (Frank 1983, Frank and Barrera 1989), and the Cell tree (Günther 1988) (see Chapter 2). This book has also presented some indexing techniques that might be incorporated: the KD2B-tree, the Sphere-tree (see Chapter 4), and the Reactive-tree (see Chapter 6). In fine-tuning the application, the proper indexing technique has to be selected.

7.5   Discussion

The Geographic Analysis Intelligence System (see Subsection 1.5.4) is an information system with typical geographic requirements. The proposed extension to the relational data model can be used as the basis for all kinds of Geographic Information Systems. In fact, the model is also useful for other pictorial applications, for example CAD/CAM systems.

I have tried to keep the extension realistic and within the concepts of the original relational data model. By 'realistic' I mean that the implementation is based on known, recently developed techniques. The implementation will be described in the next chapter. However, the extension requires a DBMS that is extendable with additional types, storage and search mechanisms.

8 Implementation of the Geographic Extension

The previous chapter revealed that GISs require a geographic extension to the relational data model. Further, the functional aspects of a geographic extension were described. This chapter describes an actual implementation of the specified extension. Most commercial GISs are closed. This means that, if certain functionality is not available, it is impossible for the users to extend or modify the system for their own purpose.

Most commercial GISs are based on a relational DBMS, such as Oracle (Oracle Corporation 1990) or Ingres (Sun Microsystems, Inc. 1987). As described in the previous chapter, one obvious drawback of the standard DBMSs is that they cannot manipulate geographic data. That is, there are no geometric attribute types (e.g., point, polylines, polygons) or operators (e.g., distance, intersection, circumference, area). Different GIS architectures try to solve these problems in their own manners. These GIS architectures are presented in the first section, each with its own advantages and disadvantages. GEO++, the system described in this chapter, is based on the extensible DBMS Postgres.

Section 8.2 gives a short introduction to Postgres. The subsequent section describes the implementation of a geometric extension to the open DBMS Postgres (Stonebraker *et al.* 1990). The user-interface of GEO++ is built on top of ET++. Section 8.4 cites the basic capabilities of the Postgres GIS front-end GEO++. The real power of GEO++ is shown in Section 8.5, in which the system is extended with user-defined types.

8.1 DBMS-based GIS Architectures

The standard DBMSs do not provide basic geographic data types so that it is impossible to store geographic data in a natural manner and to pose queries such as: 'Select all towns with more than 10,000 inhabitants that are located within 3 kilometres from a lake'. These systems also lack multi-dimensional access methods (or index mechanisms), which are required because geographic data sets are often very large.

Different DBMS-based solutions have been suggested and implemented in order to overcome these problems. Three different types of system architectures can be distinguished: dual architecture, layered architecture, and integrated architecture. The term dual architecture was first introduced in van Oosterom and Vijlbrief (1991). A similar classification was described by Bennis *et al.* (1991); they used the terms partial DBMS architecture,

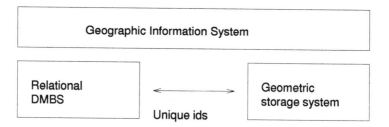

FIG. 8.1. The Dual GIS Architecture

shell architecture, and full DBMS architecture for dual architecture, layered architecture, and integrated architecture, respectively. In the next subsections, the pros and cons of these architectures are discussed. Note that it is not always easy to classify a specific system. For example, Smallworld GIS (Chance *et al.* 1990) possesses characteristics of both the layered and the integrated architecture.

8.1.1 Dual Architecture

The most common type of commercial GIS architecture is the dual one. Dual architecture GISs have a separate subsystem for storing and retrieving spatial data, while thematic information is stored in a relational DBMS. This dual architecture is not conceptually elegant and also reduces the performance. An object that has both a thematic and a spatial component has parts in both subsystems that are linked by a common identifier. In order to retrieve an object, the two subsystems have to be queried and the answer has to be composed. Fig. 8.1 illustrates this dual GIS architecture. Typical examples of GISs with dual architecture are: ARC/INFO (ESRI 1989, Morehouse 1989) from ESRI, MGE (Intergraph 1990) from Intergraph, SICAD (Schilcher 1985, Siemens Data Systems Division 1987, Singer 1991) from Siemens, and ARGIS 4GE (Unisys Corporation 1989) from Unisys.

The advantage of the dual architecture is that it is partly based on a standard DBMS and that the storage and retrieval of spatial data can be efficient. However, this method has some severe drawbacks. The existence of two storage subsystems implies that query optimization is not possible to the full extent. A relational DBMS offers transactions that are atomic, durable, and serializable. Storing data outside the relational DBMS can result in losing the transaction semantics, because the two storage managers each have their own locking protocol. The final drawback of the dual architecture is that integrity constraints can be violated. For example, an entity can still exist in the spatial storage subsystem after it has been deleted from the relational DBMS.

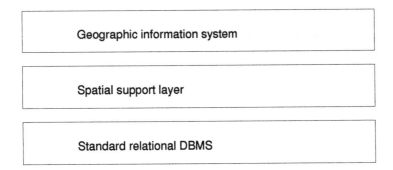

FIG. 8.2. The Layered GIS Architecture

8.1.2 Layered Architecture

Most drawbacks of the dual architecture are caused by the fact that there are two storage managers each with their own responsibilities. It is possible to store the spatial data in the pure relational data model (van Oosterom *et al.* 1989, van Roessel 1987). This implies that the support for transaction semantics and integrity constraints is restored. However, in order to fit the data into the relational model, the coherent geographic entities have to be broken into multiple parts, which are stored in separate tables. Retrieving the original geographic entities has to be done by joins of relations, making the system slower and more difficult to use.

The user may be freed from formulating difficult queries by some 'intelligent' translations in the layer on top of the standard relational database. This layer translates geographic queries (in 'GeoSQL': Geographic Structured Query Language) into standard SQL-queries and it may also implement spatial indexes. These indexes are implemented by means of auxiliary relations that contain the required index data. This makes spatial access faster, but the queries become even more complicated, because they have to use the auxiliary relations. This indirect implementation of an access method is less efficient than a direct implementation in the DBMS kernel. The layered GIS architecture is depicted in Fig. 8.2. System 9 (Pedersen and Spooner 1988) from Prime, GEOVIEW (Waugh and Healey 1987) from the University of Edinburgh, and SIRO-DBMS (Abel 1989) from CSIRO Australia are characteristic examples of layered architecture systems.

8.1.3 Integrated Architecture

The inconvenient/inefficient mapping from the user data model to the relational tables can be avoided if more attribute types and access methods are added to the system. This solution is chosen in the integrated GIS architecture. In contrast to the other two types of architectures, this one cannot be based on a standard relational DBMS. It requires an extensible (and often object-oriented) DBMS. This is illustrated in Fig. 8.3, where the spatial extension is completely embedded in the DBMS. Users may extend

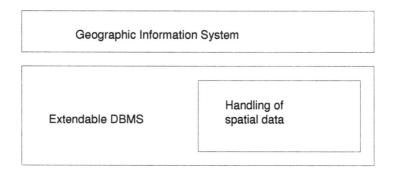

FIG. 8.3. The Integrated GIS Architecture

the DBMS with their own basic abstract data types (ADTs). An obvious drawback of the approach is that users have to implement their own types within the DBMS environment, which may be quite complicated. However, once this task has been performed, the advantages of this approach become clear. The implementation of the data model is easier, because the appropriate geometric types are now available. The formulation of spatial queries is directly supported in the extensible query language by means of added spatial operators such as distance, area, and intersection.

Perhaps the most important advantage is the good performance of these systems. The direct implementation of data types in the kernel of the DBMS is very efficient. Another facility supporting system performance is that one may provide own spatial access methods. The development of integrated GIS architectures depends on the availability of open DBMSs. Characteristic examples of integrated GIS architectures are TIGRIS (Herring 1987) from Intergraph and the research-oriented system GéoTropics (Bennis *et al.* 1990) from the University of Paris VI and IGN France. Our own system, GEO++ (van Oosterom and Vijlbrief 1991), also possesses an integrated architecture. The system-component 'Handling of spatial data' (see Fig. 8.3) is implemented in the open Postgres environment.

8.2   Description of Postgres

This section gives an introduction to the open DBMS Postgres. A more detailed functional description can be found in (Stonebraker and Rowe 1986), and several implementation decisions are discussed in (Stonebraker *et al.* 1990). The Postgres reference manual (Postgres Research Group 1991) contains all the information required to use the system. Postgres, the successor of Ingres, is a research project directed by Michael Stonebraker at the University of California, Berkeley. The characteristic new concepts in Postgres are: support for complex objects, inheritance, user extendability (with new data types, operators, and access methods), facilities to save and query historical data versions of relations, and support for rules. The latter may be used for the implementation of constraints.

The next subsections explain the features that are of interest to GISs. These features will be illustrated with some GIS examples based on the current version of Postgres (version 3.1). First, Subsection 8.2.1 describes the global architecture of Postgres. The next subsection describes the query language Postquel, and Subsection 8.2.3 gives an example with a user defined type. In Subsection 8.2.4 the spatial access method that is available within Postgres, the R-tree, is described. Postgres is an extended relational DBMS. Some object-oriented terminology is used: a relation is an object class, a tuple is an object instance, and an attribute is an instance variable.

8.2.1 The Architecture of Postgres

Postgres can be viewed as a collection of files and processes that operate on these files. The files contain the relations and data required for the access methods, that is, the B-tree or the R-tree itself. A daemon process postmaster handles the communication between the back-end (the process that does the real DBMS work and is therefore called Postgres) and the front-end or application. The postmaster starts a back-end process for each application that requests the services of Postgres. A standard Postgres application is the monitor, an alpha-numeric user-interface for Postgres. The user may state Postquel queries and the answers are displayed in a tabular format. New applications can be developed based on Postgres by using the C library functions of libpq. This library contains functions to pass the queries to the back-end and to interpret the buffers, called portals, which are used to return the results. Another way of interacting with Postgres is by using the fast path. The fast path makes it possible to call Postgres system functions directly. In this way the query language is bypassed and best performance is achieved by calling the access methods.

8.2.2 The Query Language Postquel

The query language Postquel is based on three concepts:
1. There are three kinds of data types: base types (built-in, system, and user), array types (fixed and variable length), and composite types (tuple, set of tuples, and relation).
2. The following kinds of functions are available: normal functions (C or Postquel), operators (binding of a symbol to a function), aggregate functions (count, sum, average, min, max, etc.), and inheritable functions.
3. Rules have the form: 'on condition then do action' and they are used to trigger DBMS actions.

User-defined types, with their own functions and operators, are of particular interest, because these may be used to define the geographic data types. Subsection 8.2.3 describes the user types in more detail. The database administrator may 'upgrade' user types to system types, making them available to each database created on the system.

```
[a] retrieve (dist  =
    min{distance(tower.location, "(10,15)"::POINT2)})
[b] retrieve (close =
 closest{tower.location, "(10,15)"::POINT2})
[c] retrieve (tower.name)
    where inside(town.polygon, tower.location) and
                town.name = "Amsterdam"
```

FIG. 8.4. Postquel Queries using Geographic Functions

```
#include <stdio.h>
typedef struct { double x, y; } POINT;
typedef struct { POINT centre; double radius; } CIRCLE;
        /* The internal representation */

CIRCLE *circle_in(str)
/* Convert from external to internal representation */
char *str;
{ /* Allocate new CIRCLE, parse string, return result.*/ }

char *circle_out(circle)
/* Convert from internal to external representation. */
CIRCLE *circle;
{   char *result;
    if (circle == NULL) return(NULL);

    result = (char *) palloc(60);
    (void) sprintf(result, "(%g,%g,%g)",
       circle->centre.x, circle->centre.y, circle->radius);
    return(result); }

char circle_area_greater(circle1, circle2)
CIRCLE *circle1, *circle2;
{ return(circle1->radius > circle2->radius); }
```

FIG. 8.5. A Part of the C Source Code Defining the New Type circle

The current distribution of Postgres already contains a system type example that approximates a two-dimensional variant of the extension with geographic data types that we proposed. It consists of four types: `point`, `lseg`, `path`, and `box`. The type `lseg` implements a single line segment. Polylines and polygons may be represented by `path`, which is a variable length array of `lseg`. The special case of a two-dimensional axes-parallel rectangle is represented by the type `box`. Some useful functions and operators are provided (test for overlapping boxes, test if a point lies inside a box, the distance between two points), but more are required for a really good geographic extension of the DBMS; see Subsection 7.4.2.

Some practical GIS example queries show how geographic functions might be used in Postquel; see Fig. 8.4. Query [a] is a 'minimum distance' query. The next query [b] does the same thing, but is formulated more efficiently by using the function `closest`. The last example, query [c], uses the `inside` function in order to retrieve all the names of the towers in Amsterdam.

8.2.3   Defining a User Type

The example in this subsection defines the new user type circle. This example is based on the tutorial distributed with Postgres. It is important to realise that, in Postgres, there is a difference between the internal and external representations of a type. The external representation is a character string for user input and output as used in the monitor. In the case of a circle this could be (centre_x, centre_y, radius), for example (0, 0, 1): the unit circle. The internal representation determines how the type is organized in memory, just as in the programming language C. Fig. 8.5 shows the C code that defines the internal representation of the new type circle and some C functions for it. Assume that this is stored in the file `circle.c`.

The functions `circle_in` and `circle_out` perform the conversions between external and internal representations. There is also one operator function for this type: `circle_area_greater`, which determines whether the area of the first circle is greater than the area of the second one. After compiling, which produces the object file `circle.o`, Postgres must be informed about the existence of the new type and its functions: first the conversion functions are defined (see Fig. 8.6 queries [a], [b]), then the new type [c] is defined, and finally the operator function [d] and its symbolic representation [e] are defined; a '>' sign. Now it is possible to create relations with circles in them [f], append records to them [g], [h], [i], and retrieve the circles that have an area greater than the circle (0,0,8) [j]. The result of the last query [j] is of course the circle (5,1,9).

8.2.4   The R-tree

The information in the remainder of this section is based on the beta version of the R-tree in Postgres and is provided by Mike Olson (personal communication, 1991). The performance tests with the R-tree of Postgres are carried out on DECstation 5000/200 under Ultrix 4.0. The use of the R-tree is similar to the use of the B-tree in Postgres. That is, one can use

```
[a] define C function circle_in (file = "circle.o",
       returntype = circle) arg is (char16)
[b] define C function circle_out (file = "circle.o",
       returntype = char16) arg is (circle)
[c] define type circle (internallength = 24,
    input = circle_in, output = circle_out)
[d] define C function circle_area_greater (file = "circle.o",
       returntype = bool) arg is (circle, circle)
[e] define operator > (arg1 = circle, arg2 = circle,
       procedure = circle_area_greater)
[f] create tutorial (a = circle)
[g] append tutorial (a = "(5,1,9)"::circle)
[h] append tutorial (a = "(2,2,5)"::circle)
[i] append tutorial (a = "(0,1,7)"::circle)
[j] retrieve (tutorial.all)
   where tutorial.a > "(0,0,8)"::circle
```

FIG. 8.6. The Postquel Part of Defining the New Type `circle`

```
[a] create testrel (region = box)
[b] append testrel /* append lots of tuples */
[c] retrieve (testrel.all)
    where testrel.region && "(98,20,9,10)"::box
[d] define index testind on testrel
using rtree (region box_ops)
```

FIG. 8.7. Defining an R-tree Index for the Relation `testrel`

the Postquel construct **define index** to define an index on the **region** attribute of the relation **testrel**; see Fig. 8.7 [a], [d]. The R-tree can be used with system type box and the operators shown in Table 8.1.

The relation **testrel** is populated (Fig. 8.7 [b]) with 30,000 rectangles, sides random between 0 and 1,000, and origins random distributed in three regions: 10,000 in (0, 0, 10,000, 10,000), 10,000 in (30,000, 10,000, 50,000, 30,000), and 10,000 in (0, 0, 50,000, 50,000). This data set is chosen because it is representative for map data: objects of different sizes and a population density that is not constant over the whole region. Fig. 8.7 [c] shows a rectangle overlap query. This is an important type of query, because it is used to generate maps on the rectangular screen. A point query, used for implementing a 'pick' operation, can be formulated by taking a box with equal diagonal corner points.

A restrictive spatial query, which retrieves up to 100 objects without the R-tree, takes about 85 seconds using a sequential scan. Building an R-tree index on the test relation with 30,000 objects takes about 35 minutes.

Table 8.1. Box Operators Supported by the Postgres
R-tree

OPERATOR	MEANING
a << b	box a is strictly left of box b
a &< b	a is left of b, or overlaps b,
	but does not extend to the right of b
a && b	a overlaps b
a &> b	a is right of b, or overlaps b,
	but does not extend to the left of b
a >> b	a is strictly right of b
a @ b	a is contained by b
a ~ b	a contains b
a ~= b	a and b are the same box

However, now the same spatial queries run typically a few hundred times
faster using the index scan. As the size of the relation grows, the gain of
the index-scan will become larger and larger compared with the sequential
scan.

The size of the file that contains the `testrel` is 3.0 Mb. The size of the
file that contains the index `testind` is 6.7 Mb. This may seem a lot in
comparison with the relation `testrel`, but, in the case of GIS application
with tuples that have polygon attributes varying from 10 to 1,000 points,
the overhead of the R-tree is quite acceptable.

8.3   Geometric Data Types

The first step in the implementation of a GIS with an integrated architecture is to extend the DBMS with geometric data types and operators.
In our case, Postgres itself has some geometric capabilities. It comes with
the R-tree (Guttman 1984) spatial index structure and some geometric
types; see Subsection 8.2.2. However, we implemented both two- and three-
dimensional geometric data types with a more complete set of associated
functions and operators. Therefore, the type name ends with a number
that indicates the dimension. The 2D data types are: `POINT2`, `POLYLINE2`,
and `POLYGON2`. Instead of using the Postgres system type `path`, we want
to reflect the semantic difference between polyline and polygon by pro-
viding separate data types for them. These types have to be supported
with more geometric functions (Mercera 1991). Function names reflect the
purpose, the dimension, and the operand types. For example, the function `Equal2PntPnt` checks if two two-dimensional points are equal. The
functions are classified into groups, each illustrated with a few examples
(operator symbols are indicated within square brackets):

Return Boolean values:

`Equal2PntPnt [==]`

`On2PntPln [-->]`

`Contain2PgnPgn [-->]` (operator overloading!)

`Overlap2PlnPgn [&&]`

Return atomic geometric objects:

`GravCentre2Pgn [@]`

`MinBoundRec2Pgn [|=|]`

`ConvexHull2Pnts [~]` (Pnts indicates a variable length array of `POINT2`s)

Return complex geometric objects:

`Inter2PgnPgn [^]`

`Voronoi2Pnts [<|]`

`Delaunay2Pnts [|>]` as described in Preparata and Shamos (1985)

Return scalar values:

`Distance2PntPnt [<->]`

`MinDist2PgnPgn [<->]`

`Length2Pln [--]`

`Area2Pgn [*]`

The functions from the last category can be performed either on the flat 2D surface or on the curved surface of the earth. The dimension in the function name is indicated as 2E and all operator symbols start with $\sim$, e.g. `Length2EPln [~--]`.

8.4   Basic GEO++ Capabilities

GEO++ is a general-purpose GIS front-end for Postgres. The system has a 'direct manipulation user-interface'. GEO++ can be used to implement real-world GIS systems, and to experiment with the user-interface and various data structures and storage techniques. Some of the expected applications are electronic sea maps and various Command and Control (C2) systems. The current prototype system is written in C++ (Stroustrup 1986) and uses the ET++ (Weinand *et al.* 1988, 1989) class library. The screendump (Fig. 8.8) shows a part of the user-interface of GEO++. The interaction with GEO++ is based on the following capabilities:

1. Alpha-numeric viewing of Postgres relations.

2. Addition or editing of Postgres tuples in viewed relations.

3. Graphically displaying the result of queries on map. Point attributes are shown as coloured icons. These icons and their colours are also specified in GEO++ meta relations or are taken from the displayed relation. Polylines and polygons can also be shown. The width (polylines) and colours are specified in the same way as for the point attributes.

4. Possession of its own meta relations, for instance, a relation that describes the regular expression that each entered Postquel-type constant must match. This is used by GEO++ to ensure the construction of correct 'where' clauses.

5. Graphically composing the 'where' clause restriction for the viewed re-

lation in a tree format that represents the structure of this expression. The graphical editor inspects the Postgres meta relations (system catalogues) which contain information about the existing operators. In this way, users have access to all defined operators and types, including types they have defined themselves.

6. Facility enabling the composer to check the types of arguments to operators and forbid clauses that would result in syntactical errors. Composed query-trees can be saved and reloaded for different relations.

7. Labelling geographic entities on the map. The labels are formed by the attribute name (optional) and the current value.

8. Facility enabling users to shuffle the drawing orders of the displayed queries or temporarily hide them. Also, parameters of the 'where' clauses can easily be modified.

9. Picking of displayed objects. This shows the corresponding Postgres tuple.

10. Editing and creating point, polyline, and polygon attributes on the displayed map.

11. Zooming and panning of the displayed map and of a second map display which shows the context of the main map. This is supported by a spatial access method: the R-tree.

12. Automatic redraw of changed relations by means of Postgres asynchronous rules (triggers). This capability allows the creation of dynamic map displays with moving objects.

13. The possibility of customizing the GEO++ system by hiding features and options of the system that are not needed for a specific application or adding special functions or icons by editing metadata in the Postgres database.

14. Annotating a map with text, lines, polylines, symbols, etc., like a normal drawing program. Multiple annotations (from different users) may be displayed simultaneously, thereby allowing communication about map contents between two or more remote users.

In the Postgres layer, first one specifies the database table and the attributes that should be retrieved. Then the selection criteria can be specified by building a tree which represents the 'where' clause of the query. The resulting tree is the graphic representation of the parse tree of a 'where' clause in the Postquel query language. The operations that the user can apply to the nodes in this tree are the productions in the context-free grammar describing a Postquel 'where' clause, although the end-user is probably not (and should not be) aware of this underlying principle. This guarantees that all possible 'where' clauses can be specified. This graphic tree building has two advantages:

1. The graphic tree representation makes inherently complex Boolean queries easier to understand (parse) for the end-user.

2. It is not possible to formulate queries that result in a syntax error. The system checks the parameter types of functions and operators and guides the user by the selection of the actual parameters. For example, if

the user chooses a function like `Distance2PntPnt`, then he can only select table attributes that are of the correct type (`POINT2`).

Most errors made by the users, however, are semantic and not syntactic. This cognitive aspect warrants further research. The productions (rewriting rules) are:

- Choosing a function or operator, which is used for implementing the restriction.
- Choosing a table attribute, which is an operand for a selected function or operator. This may be an attribute of another table (implementing joins) or of a previously composed query.
- Choosing a constant from a range of types (`bool`, `int4`, `POINT2`, `POLY-LINE2`, `POLYGON2`, `text`, etc.) as operand. The available types are retrieved from the Postgres system tables.
- Choosing a Boolean operator: `and`, `or`, and `not`. This enables the user to create more complex queries.

An example using the relations, `o1` and `o2`, which both have attributes height and location, will illustrate the interactive formulation of a 'where' clause. The query in Fig. 8.8, which retrieves (and displays) all objects of relation `o1` that have a height less than 20 and whose distance from at least one of the objects in relation `o2` is less than 100 kilometres, could be composed by the following sequence:

1. Choose `operator < (float4,float4)`
2. Choose `attribute o1.height`
3. Choose `constant` and enter 20
4. Choose `Boolean AND`
5. Choose `operator < (float8,float8)`
6. Choose `operator <-> (POINT2,POINT2)`
7. Choose `attribute o1.loc`
8. Choose `attribute o2.loc`
9. Choose `constant` and enter 100.

Note that constants are automatically cast to the correct type, that is, `float4` in line 3 and `float8` in line 9. However, attributes can be selected only if they are of the correct type, that is, `float4` in line 2 and `POINT2` in lines 7 and 8.

Of course, the same query could be composed by applying the productions in another order. For example, start with `operator AND`, choose `operator < (float8,float8)`, etc. The only restriction in the current version is that the operator has to be selected before the operands can be chosen. So one has to know the type of the operands.

8.5   Extending GEO++

Just as Postgres is extensible by defining new abstract data types and coding their implementation in the programming language C, GEO++ is extensible by defining new `QueryShapes` and coding them in C++. The

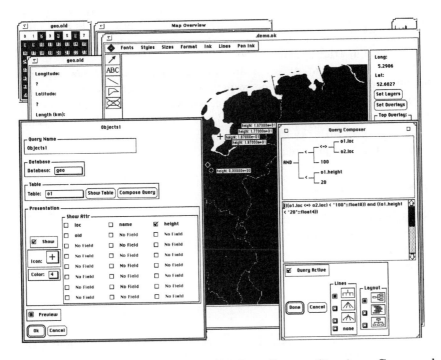

FIG. 8.8. The User-Interface of the GEO++ System, Showing a Composed Query

advantage of C++ is that one can inherit most of the functionality of the parent `QueryShape`.

The new `QueryShapes` must implement a `draw` method and a `distance` method that are used for drawing the `QueryShapes` on the map and for the 'pick' operation. Typically, a GEO++ `QueryShape` matches a user-defined geometric type in Postgres. The geometric operations are implemented in Postgres and the visualization is implemented by new user-defined GEO++ `QueryShapes`.

By means of extending the GEO++ system, one has almost unlimited flexibility in the way data are stored in Postgres and displayed with GEO++. Below, we shall see how `QueryShapes` that display data of our own set of geometric types (Section 8.3) are entered into a GEO++ meta relation. This is required because GEO++ was originally based on the standard geometric types of Postgres.

We have implemented, for example, a `Pln2Shape`, a `Pgn2Shape`, and a `Pnt2Shape` that visualize the corresponding new Postgres geometric types. GEO++ consults the meta relation `geo_dyninfo` when displaying relations. If we have a relation `bordersmap` that has a `POLYLINE2` attribute which has to be displayed with the `Pln2Shape`, then we must enter these data with the following Postquel command:

```
append geo_dyninfo (
  relname="bordersmap",
  dynfunc="make_Pln2_shape",
  dynfile="MapShape.o",
  relattr="plndata,colour,width",
  bboxattr="bbox"
)
```

This specifies that:

- The tuples in the relation `bordersmap` are displayed by calling the C++ function `make_Pln2_shape` defined in `MapShape.o`.
- The function is called with the attributes `plndata`, `colour`, and `width` as arguments.
- The objects in `bordersmap` have a bounding box that is stored in the attribute `bbox`.

The 'open' character of GEO++ is demonstrated by the implementation of a `QueryShape` that can be used to display raster-type data, thereby creating a hybrid system (Vijlbrief and van Oosterom 1992). Raster data are stored in variable size clusters (e.g., 32×32 or 64×64) that store rectangular parts of the raster-type data. This method takes full advantage of the R-tree access method. The values at the edges of a cluster could be duplicated from the neighbour cluster of each edge. This allows efficient implementation of operations that need access to the neighbours of each raster cell.

8.6   Discussion

Compared with many commercial GISs, the current version of the GEO++ system contains less functionality. However, the sound basis of an extensible DBMS and GIS allows the users to incorporate the functionality they require. Not only completion of existing capabilities, but also new areas of functionality may be added, for example raster data types or 3D data types, together with a set of appropriate operations, and visualization techniques.

Postgres offers several mechanisms for developing advanced GISs that have not yet been exploited in GEO++ system. For example, Postgres offers historic data and versions of relations. There are types of GISs in which this plays an important role: C2 systems, GISs monitoring of environment, or GISs visualizing census data. It is obvious that these kinds of applications will benefit from the automatic storage of historical data.

Applications that require geographic data at multiple scales are another example where the novel mechanisms of Postgres might offer solutions. We are developing a system that avoids storing redundant data, i.e. the storing of a separate map for each scale. This system might benefit from a combination of techniques:

- The Reactive-tree (van Oosterom 1991) can be implemented within Postgres.
- Rules can be used to derive small-scale maps of large-scale databases (Müller 1990*b*).
- Functions within Postgres are useful for the implementation of procedural map generalization techniques. For example, associated with a polyline or polygon is a line generalization algorithm to reduce the number of points used when working with small-scale maps.
- Composite type attributes (relation, (set of) tuple) can be used for the multi-scale representation of a single object. These attributes allow references to other tables, which describe the refinement of objects in a larger-scale map (Chang and Kunii 1981).

Another important research area deals with the cognitive (user-interface) aspects of a GIS. Tests with real users are necessary to determine what a 'good' graphic interface to GISs should look like. This covers both the input and the ouput side of the interface. On the input side, it is clear that the direct use of Postquel by end-users is not optimal. On the output side, the users will benefit from an enhanced cartographic interface: more context information displayed (legend, north, scale), more types of projections, **QueryShapes** to produce thematic maps (choropleths, prism-maps, maps with pie-charts, etc.).

9 The Object-Oriented Approach and Procol

The applicability of the object-oriented (OO) approach to GISs is analysed in this chapter. In software engineering, the OO approach as a design model has been proved to produce quality software. GISs might also benefit from the OO approach. However, a GIS also imposes special (e.g. spatial) requirements, inclusion of which in the OO model has to be investigated. The proposed solution attemps to meet these special requirements by incorporating two data structures: the R-tree and the BLG-tree (see Part I).

The OO approach can be useful during several phases of the development of a GIS: information and requirements analysis, system design, and implementation. The approach also offers good tools for the database and user-interface subsystems of an information system. The interest in object-oriented GISs appears to be growing (Clementini *et al.* 1990, Clementini and Felice 1990, Egenhofer and Frank 1989*b*, Gahegan and Roberts 1988, Mohan and Kashyap 1988). The OO approach has proved a good method for these purposes in other information and software systems. It leads to quality software which is extendable and reusable (Meyer 1988). The reuse of software itself tends to improve the quality because reuse implies testing and, if necessary, debugging. The OO approach also results in good user-interfaces, including both visual and non-visual aspects. The explanation for the latter is that if the objects are well chosen the user can easily form a mental model of the system.

In some sections of this chapter the OO approach will be illustrated by reference to the presentation of census data (see Subsection 1.5.2). Each such section consists of two parts: one describing the principles of the OO approach and one dealing with the case study. The latter will be illustrated by objects described in the language Procol, a Protocol-Constrained Concurrent Object-Oriented Language (van den Bos 1989*a*, 1989*b*, van den Bos and Laffra 1989, 1990, 1991). The next section summarizes the aspects of Procol that are relevant to this book.

In this book the emphasis is on the design of GISs and their databases. However, design must be preceded by information analysis to obtain the requirements. Section 9.2 contains some notes on object-oriented information analysis for a GIS. The design aspects concerning the database and related search problems are discussed in Section 9.3. Section 9.4 deals with avoiding data redundancy and also introduces topology. This is accomplished by the CHAIN object. Geographic generalization is the topic of Section 9.5. The term 'generalization' is intended in its cartographic sense and not in the sense often used by computer scientists in the inheritance hierarchy of the data or object model. The BLG-tree, a data structure for line generalization, has already been discussed in depth in Chapter 5. Section 9.5 concerns other generalization techniques in the context of the OO approach.

9.1 Procol Overview

This section gives a short summary of Procol. A detailed functional description of Procol can be found in van den Bos (1989a) and van den Bos and Laffra (1990, 1991), and some implementation aspects are given in van den Bos and Laffra (1989). Procol is a parallel object-oriented language supporting a distributed, incremental, and dynamic object environment. It also supports parallel delegation, a dynamic replacement for the conventional concept of inheritance. It is a superset of C (Kernighan and Ritchie 1978). A Procol object definition consists of the following sections:

OBJ	*Name Attributes*
Description	natural language description
Protocol	(sender-message-action)-expression
Declare	local type definitions, data and procedures
Init	section executed once at creation
Cleanup	section executed once at deletion
Actions	definition of public methods
IntActions	definition of public interrupt methods
EndOBJ	*Name*

Internally an object executes sequentially. Externally, in relation to other objects, objects run in parallel, as long as they are not engaged in communication. The channelling of information from one object to another is accomplished by message exchange. An object offers a number of services (methods) to other objects. In Procol the services are specified in the **Actions** section. The name of an action is known externally. The body of an action may include, besides standard C-code, **send** and **request** type communication primitives.

9.1.1 Communication

For a **send** the sender of the message waits until the message has been accepted by an intended receiver. The potential receiver is likewise suspended until it acquires the required message. Receipt of the message consists of copying the values of the message's components to variables local to the object. Immediately after receipt of the message both the sender and the receiver resume execution. Any processing of the received message is done after the sender has been released.

The **request** is comparable with the type of communication in Ada. Sender and receiver are bound until a result is returned. However, the result is returned by means of a **send**. This allows for early return, with postprocessing in the server possible. In fact, only the client knows whether an action in the server is triggered by a **request** or by a **send**, another contribution to information hiding. Sending a message to an object via a **send** uses the syntax:

```
TargetObject.ActionName(msg)
```

`ActionName` is the name of the action in `TargetObject` to which message `msg` is sent for processing. The **request** uses a generalization of the **send** syntax:

```
TargetObject.ActionName(msg) -> (result)
```

The values returned by the **request** are deposited in the variables indicated in the list **result**. The list must originate from a single **send** in the server. `TargetObject` can be the following:

- a variable containing the identity of a particular object;
- one of the constants **Creator** or **Sender**;
- the name of an object type, indicating any instance of that type;
- **ANY**, indicating any instance from the universe of object types.

The first two cases correspond to 1–1 communication mapping, the latter two with 1–n mapping. The constant **Creator** indicates the identity of the creator of this object. **Sender** is the primitive yielding the name of the object that sent the latest message accepted by the object issuing this primitive.

The body of an action can further contain the Procol primitives **new**, which creates a new instance of an object and executes its **Init** section, and **del**, which deletes an existing instance of an object after executing its **Cleanup** section. Deletion is allowed only by the creator of the object, the identity of which is acquired by the Procol primitive **Creator**. The bodies of the **Init** and **Cleanup** sections have the same form as the bodies of the actions.

Only one action per object can be executed at a time. Normally the object processing an action first completes it before it can receive any new message. However, interrupt actions, if present, may temporarily interrupt an ongoing (non-interrupt) action; see Section 9.1.3.

9.1.2   Delegation

An action may also contain a *delegation* to another object. This is indicated by the syntax:

```
@DelegateObject.ActionName(msg)
```

in which `msg` is not necessarily identical to the message received by the delegating action. For the client delegation is transparent; its server remains the object that it communicated with initially. Analogously, for the delegate object the client (**Sender**) remains the original requester. Technically, delegation is a **send** in which the sending object is disguised as the original client. The implication is that as soon as the message has been transferred control returns to the delegating action. Multiple levels of delegation (nesting) may occur.

9.1.3 Interrupt Actions

Interrupt actions (in the **IntActions** section) are public actions that may be accessed via a message, just normal actions. However, when a normal action is in progress, the interrupt action in the same object has priority. The execution of the normal action is suspended and the interrupt action is executed. Once the latter is completed, execution resumes at the point of suspension in the normal action.

To be allowed access, interrupts have to appear in a compound protocol (see Subsection 9.1.4). Interrupt actions cannot interrupt themselves or other interrupt actions. Through the protocol, interrupts may be masked by any action using the device of guards.

9.1.4 Protocol

Procol constrains possible communications by controlling access to the object's actions, by a **Protocol** section in each object. Protocols also define a temporal order on input (messages from other objects to this object), thus acting as a major aid towards structuring input patterns. The protocol has the form of an expression over *interaction* terms, the counterparts of **send statements** and **request** statements. During the life of an object this expression is matched against actual communications. The protocol is repeated. An interaction term couples the reception of a message to an action. Its form is:

```
SourceObject(msg) -> ActionName
```

The semantics is that, upon receipt of message `msg` from `SourceObject`, `ActionName` will be executed. `SourceObject` can be specified as: object instance, object type (class), and **ANY** object. In the receiving object, the Procol primitive **Sender** contains the identification of the communication partner. Expressions over interaction terms are constructed with the following four operators (in this table E and F themselves stand for interaction terms or for expressions):

E + F	selection:	E or F is selected
E ; F	sequence:	E is followed by F
E *	repetition:	Zero of more times E
phi:E	guard:	E only if phi is true

I shall illustrate the expression operators with a few simple examples. Let us assume that we specify the protocol in a server object S with actions `actionA` and `actionB` triggered by messages `msg1` and `msg2`. Clients A and B exist also. Then the following protocol in S

```
ANY(msg1) -> actionA + ANY(msg2) -> actionB
```

is equivalent to no protocol as far as access control is concerned, because it specifies that any client may send `msg1` to trigger `actionA`, or `msg2` to trigger `actionB`. The protocol does however play a role as interface specification for the outside world. The following protocol in S

```
A(msg1) -> actionA + B(msg2) -> actionB
```

specifies that only clients A and B have access to object S; A can send a message that triggers `actionA` and B can send a message that triggers `actionB`, with the exclusion of any other access patterns. The following protocol in S

```
A(msg1) -> actionA ; A(msg2) -> actionB
```

specifies that only client A is allowed access to object S; furthermore, A first has to send a message S.`actionA(msg1)` to trigger `actionA` before it can send a message S.`actionB(msg2)` to trigger `actionB`. Since the protocol repeats, this doublet of messages repeats as well. Although the protocol as a whole repeats, it is sometimes necessary to repeat a sub-expression in it as well. The following protocol accomplishes that:

```
A(msg1) -> actionA * ; B(msg2) -> actionB
```

It specifies that client A may trigger `actionA` one or more times or not at all, and that such a repetition always must be concluded by B triggering `actionB` once. Hence the interaction with object B serves as a terminator in this example.

9.1.5   Guards

To extend the power of the (so far regular) expressions, predicates in the form of guards may be used. A guard precedes an interaction term or an interaction expression. It is evaluated before any actual communication, as specified in the interaction expression subjected to the guard, takes place. A guards evaluation yields either true or false. It can be used to receive a message conditionally, and thus to execute the corresponding action conditionally. If more than one guard occurs in a selection, they are all evaluated. In general, a guard is a function without side-effects, testing some aspect of the state of the object.

The form of a guard is a test expression followed by a colon (:). As an interaction expression operator, the guard operator has the highest precedence. An example is a protocol with guards opening or closing access to actions that insert (`Put`) a symbol `sym` in a buffer, or fetch (`Get`) a symbol from the same buffer:

```
count <= size: ANY(sym) -> Put
+
count > 0:     ANY -> Get
```

in which **size** is the size of the buffer, and **count** is the number of buffer slots occupied. These two variables are set by actions **Put** and **Get**. The upshot is that no client may insert a symbol when the buffer is full, and no fetching is possible until at least one symbol is present. Clients that issue requests that cannot be honoured remain pending until they can be satisfied as a result of a change of state of the object that makes the corresponding guard true. Any action that completes causes a re-evaluation of the guards.

9.1.6 Compound Protocols

A protocol may be compound. In that case it consists of two or more (simple) protocols separated by the ‖ operator. All constituent protocols are simultaneously active. As an example, take an object that performs interactions **OpenR**, **Read**, **CloseR**, **OpenW**, **Write**, and **CloseW**, for reading and writing files. By using the following compound protocol:

```
OpenR ; Read* ; CloseR
||
OpenW ; Write* ; CloseW
```

read and write operations may occur interleaved, as long as each of them maintains the usual open, read/write, close temporal order.

9.2 Finding the Objects: Information Analysis

9.2.1 Some Principles

The problem area must be analysed, just as in other cases of system development. The application area must be well understood and the task of the new system within it must be clear. This results in a set of requirements for the new system. The functional requirements and the data affected must be defined in terms of objects' types. This means that during the information analysis special attention must be paid to what might possibly be the objects in the target system (van den Bos 1989*b*). There is only one thing that determines the objects of the functional requirements: the entities the user is working with and thinking about. These objects are representations of entities in the real world. After we identify an object, we ask ourselves what we want this object to do for us (services) in the context of our application. If we cannot come up with useful services, this entity is not worthy of being an object in our system.

The resulting object should be reasonably coherent and autonomous: strong internal coupling, weak external coupling. Stated in a simple manner, this means that all components (state variables and functions) within the object must be related to each other and that the use of other objects is restricted to a few well-defined methods, which specify the object interface. An object type (class) is described by a set of attributes and their associated operations (also called actions or methods). The data can be accessed only from the outside by means of the services. In fact, this property is called information hiding, and is the basis for the quality of OO software.

In a GIS there must be at least one object type with a geometric attribute. The type of this geometric attribute fixes the basic form of the graphic representation of the object. There are two categories of geometric attributes: raster and vector. The vector type can be subdivided into point, polyline, and polygon. In an interactive information system the interface must be user-friendly and the required operations must also be performed fast (interactivity). In the case of a GIS this implies that special attention has to be paid to the spatial organization of the objects. Another very important requirement for an interactive GIS is that the user should be able to look at the data (for example thematic maps) at several levels of detail; see Section 1.7.

9.2.2   The Presentation of Census Data in Procol

In the presentation of census data described in Subsection 1.5.2, the five different types of administrative unit were identified as the most important objects: municipalities, economic geographic regions, nodal regions, corop regions, and provinces. The operations described there (display, report, and update) were quite general. They have to be present for all five object types, with exception of the 'border cases'. For instance, it is impossible for a municipality to collect data from objects lower in the hierarchy because it is already the lowest object type in the hierarchy. A part of the Procol code describing the object type Economic Geographic Region (EGR) is given below. In this example the object types, MUN (municipality) and NODAL, and the data type polygon are already known.

```
/* somewhere:
   OBJ NODAL ...
   OBJ MUN ...  */
------------------------------------------------------------
OBJ         EGR (geom, name, nrof,..., parent)
            /* Economic Geographic Region */
            polygon    geom;
            string     name;
            int        nrof,...; /* census data */
            NODAL      parent;
Description Part of the object EGR
Protocol    not EGRComplete:   MUN() -> AddMUN +
            EGRComplete:     ( MUN() -> DelMUN +
                               ANY() -> Display +
                               ANY() -> RetrieveData +
                               ANY() -> ChangeData    )
Declare     struct list {
                 MUN           child;
                 struct list   *next;
            } *first, *curr;
            void Aggregate (...) {...};
            boolean EGRComplete;
```

```
Init            parent.AddEGR;
                first = curr = NULL;
                EGRComplete = FALSE;
Cleanup         parent.DelEGR;
Actions         AddMUN = {...}
                DelMUN = {...}
                Display = {...}
                RetrieveData = {...}
                ChangeData = {...}
EndOBJ          EGR.
```

The other object types are similar. However, the top-level objects (type Province) do not have an attribute **parent**. The top-level objects are created first, then the second-level objects, and so on. The **Init** section of EGR specifies that when a new instance is created an **AddEGR** message is sent to its **parent**. This **parent**, a nodal region, then adds the EGR to its list of EGRs. Note that the **parent** is not the same object as the Procol primitive **Creator**. In a normal situation an object has received all the **Add...** messages from its children before it receives any other message type. This is ensured in the **Protocol** section by the use of the guard construction. The EGR sets its variable EGRComplete to TRUE, when all municipalities that belong to the EGR have sent their registration message, otherwise it has the value FALSE. EGRComplete, which is initially FALSE, is used to control the access to parts of the protocol. EGRComplete can be set in the body of the AddMUN and DelMUN actions.

9.3   Object Management and Search Problems

9.3.1   Some Principles

In the previous section we saw that in general an object consists of a data-part and an operation-part. Normally, the object instances are present only when the program is executing. The data must be loaded into the (new) objects when the program is initialized from a file or a database system. Just before the program stops, the data must be saved again. This is an inconvenient method, especially in the case of GISs in which there are huge amounts of data that are not all needed during every session. An OO step in the right direction is to store the objects themselves, including the data (attributes, local variables, and protocol status). This can be compared with making a core dump of a single object. When an object is saved, a 'snapshot' of the object instance is made. Changes made after the 'save' operation are not transferred to this snapshot.

The suggestion of storing the objects themselves is not as simple as one might expect. This is because objects usually contain references (in attributes or local variables) to other objects. A reference to an object is an id (identification of the right type), which the operating system assigned to that object when it was created with the Procol primitive **new**. These ids are stored in attributes and variables of the right type and also in the Procol primitives **Creator** and **Sender**. In some situations it is useless to save an object without also saving the related objects.

A better solution to this problem is to have some kind of object management system (OMS) that takes care of persistent objects. If during the execution of a program an object wants to communicate with an object that is not yet present, the OMS tries to find this object and load it. It is the responsibility of the OMS to keep the references in the object system consistent. The function of the Procol sections **Init** and **Cleanup** must be reviewed for persistent objects. There should probably be a mechanism to indicate in which persistent objects one is interested. This limits the scope, so the chance that the OMS finds the wrong communication partner decreases. Chapter 10 describes the OMS of the Procol extension that includes persistent objects.

The objects as specified in the previous section are unsuited for search operations. If we want to solve the query: 'How many inhabitants has the municipality with the name Leiden?' we have to look at all the instances of the MUN object type until we have found the right one. This is an $O(n)$-algorithm. However, this problem can be solved with an $O(log(n))$-algorithm, if a binary search is used (Knuth 1973). In a relational database, efficient search is implemented by a B-tree (Bayer and McCreight 1973) for the primary key and for all other keys on which an index is put. The B-tree has many useful properties; for example, it stays balanced under updates, it is adapted to paging (multiway branching instead of binary), and it has a high occupancy rate.

The searching problem also applies to the geometric data. If no spatial structure is used, then queries such as 'Give all municipalities within rectangle X' are hard to answer. A spatial data structure that is suited for the OO environment is the R-tree (Guttman 1984). This is because the R-tree already deals with objects; it only adds a minimal bounding rectangle (MBR) and then it tries to group the MBRs that lie close to each other. This grouping process is reflected in a tree structure, which in turn may be used for searching.

9.3.2   The Case Study

In Procol, binary trees can be implemented in two ways. The first method stores the whole tree in one search object. The second stores each node of the tree in a separate instance of the search object. The latter introduces overhead by creating a lot of search objects (nodes). However, it has the advantage of being suited for parallel processing in Procol because the search objects can run on parallel processors. This is useful for range queries such as: 'Give all municipalities with more than 10,000 and less than 20,000 inhabitants'.

There must be a separate search tree for each attribute (for which efficient searches are required) on each detail level. This must be made clear to the OMS. So, if we are interested in the number of inhabitants, we have to build search trees on the detail levels of municipalities, economic geographic regions, nodal regions, corop regions, and provinces. A more sophisticated approach would be based on a reactive data structure; see Chapter 6. However, in this chapter each detail level gets its own search tree.

It is possible that the value of an attribute changes after the search tree has been created for that attribute. In that case the tree may become incorrect or inconsistent. This can be solved by sending an (implicit) message to the search tree object(s) in the OMS just after the attribute has changed. Upon receipt of this message, the search tree adjusts itself.

9.4 Topology and Avoiding Redundancy

9.4.1 Some Principles

An additional design criterion is that redundant data storage should be avoided. This applies to both geometric and non-geometric data. There is one exception to this requirement: attributes which are difficult to aggregate and which are often required may be stored redundantly. So high efficiency dominates over low redundancy. In this situation special protocols and actions must guarantee the consistency of the data.

9.4.2 The Case Study

In our case study the census data are only stored in the administrative units in which the data were collected. The collection and aggregation of census data is very simple. Just retrieve the data from the children and add them.

The type polygon as used in the attribute geom of administrative units is not yet specified. If it is a list of coordinates (x, y), then the attribute geom introduces redundancy in several ways. At a fixed detail level, neighbours have common borders. This implies that all coordinates are stored at least twice. There is also redundancy in the attribute geom between the detail levels. A border of a province is also a border of corop regions, and so on. It is possible that the same coordinate is stored up to ten times.

In principle, this is an undesirable situation. We introduce a new object type named CHAIN. A chain is a part of the border of a municipality that contains a node only at the start and end-point. A node is a point where three or more municipalities meet (see Fig. 7.4 on p. 107). These are the normal definitions in the topological data model (Boudriault 1987, Peucker and Chrisman 1975). The attributes of the CHAIN object are: a sequence (array) of coordinates, and references to a left and a right municipality. A part of the Procol code describing the object type CHAIN is given below.

```
/* somewhere:
   typedef struct { float x, y;} Point;
   typedef enum { LEFT, RIGHT } Position;
   typedef ... BLG; */
```

```
-----------------------------------------------------------------
OBJ           CHAIN(nop, points, left, right)
              int        nop;          /* number of points */
              Point      *points;
              MUN        left, right;
Description   Part of the object CHAIN
Protocol      ...

Declare       BLG tree, Build(...) {...};
Init          left.AddCHAIN(LEFT,right);
              right.AddCHAIN(RIGHT,left);
              tree = Build(nop, points);
Actions       ...

EndOBJ        CHAIN.
-----------------------------------------------------------------
OBJ           MUN(name,..., parent)
              string       name;
              EGR          parent;
Description   Part of the object MUN
Protocol      /* Lowest object in hierarchy */
              not BoundaryComplete:
                  CHAIN(pos,neighbour) -> AddCHAIN
              + BoundaryComplete:
                  (ANY() -> Display +
                  ANY() -> RetrieveData +
                  ANY() -> ChangeData )
Declare       struct list {
                  CHAIN          chain;
                  struct list    *next;
              } *first, *curr;
              MUN            neighbour;
              Position       pos;
              boolean        BoundaryComplete;
Init          parent.AddMUN;
              BoundaryComplete = FALSE;
              first = curr = NULL;
Actions       AddCHAIN = {
                  ... /* add to list */
                  parent.UpCHAIN(Sender,neighbour);
              }
              Display = {...}
              RetrieveData = {...}
              ChangeData = {...}
EndOBJ        MUN.
-----------------------------------------------------------------
```

The attribute `geom` is removed from the five administrative units. Instead of this attribute there is a local variable which contains a list of (references to) chains. Initially, the list is empty because the administrative units are created before the chains. The `Display` action of an administrative unit should not be called before all chains are created. This can be controlled with a proper protocol.

The chains are created after the administrative units have been created. In the **Init** section of the `CHAIN` object a message is sent to the `AddCHAIN` action of the municipality left. This message says that the municipality is to the left of the chain and that its neighbour is the municipality right. A similar message is sent to the municipality right. After these messages have been sent, the BLG-tree (see Chapter 5) of the chain is built and assigned to the local variable `tree`.

The municipality receiving an `AddCHAIN` message knows which chain sent the message. This chain is added to the list of chains. Finally, the municipality sends a message to the `UpCHAIN` action of its father (an `EGR` object), with parameters: the chain (contained in the Procol primitive **Sender**), and the neighbour municipality. The `EGR` object also has a list of chains. Upon receipt of the message from the municipality, it checks whether the neighbour is in the list of municipalities. If it is, nothing is done (internal boundary); if not, the chain is added to the list of chains and again a message is sent to the `UpCHAIN` action of its father (a `NODAL` object). This process is repeated on all detail levels if necessary.

Notice that, with the new object `CHAIN`, we have also introduced a topological structure (Boudriault 1987). The chain can be used to determine the neighbours of an administrative unit. This is possible at each detail level.

Another interesting issue concerning redundancy is whether the area of an administrative unit should be stored. The area can be derived from the polygon (list of chains), but this is a non-trivial operation. The area is useful not only as such, but also in the display of relative information. For example, the population density is the number of inhabitants divided by the area. Because the area is so often needed and is difficult to compute, we allow some redundancy here. To guarantee consistency, a border must send a message to the administrative units if it changes. The administrative units will then update their area values.

9.5 More on Generalization

Geographic generalization, that is, the process of transforming large-scale maps (with detailed data) into small-scale maps in the design of a GIS, is supported in several places in this book:

- Things can be removed. For example, a border between two municipalities in the same province may not be drawn when displaying a small-scale map.

- An object can be drawn with less resolution (simplified) by using the BLG-tree. For example, a (part of the) border of a country may be drawn with only 5 points on a small-scale map; the same part would be drawn with 50 points on a large-scale map.

We will shortly examine some of the other generalization techniques, and comment on whether they can be supported by the OO approach to GIS. This does not imply that a certain part of the generalization can be automated, only that the result of the generalization process is reflected (stored) in the system. The major generalization techniques are as follows (see also Sections 2.3, 6.3, and 6.5):

1. The geometric representation of an object is changed. A city on a small-scale map is represented by a point (marker) and on a large-scale map by a polygon. Though not described in this chapter, it is possible to store multiple representations of an entity in one object. When this object is displayed, which representation is used is decided on the basis of the current scale and resolution of the output device. 2. An important object is exaggerated on a small-scale map; e.g., an important road is drawn wider than it actually is. It is not very difficult to implement this type of generalization if the importance of the object is stored. A simple adaptation of the Display action should do the trick (with a procedure call to 'set line width'). 3. An object is slightly displaced on a small-scale map because otherwise it would coincide with another object. For example, a city is moved because a road is exaggerated. This could be annoying. The displacement can be stored, but this depends on the scale. A solution might be a procedure that determines the displacement, but this is a very difficult process, which involves the subjective human taste or preferences. These may be simulated to a certain extent by an expert system, but such a process is not within the scope of this book. Another solution is to store a number of displacements, each of which is valid for a certain range of the scale. The number of ranges (and their sizes) varies from object to object. 4. Two or more objects are grouped and represented by one graphic primitive. For example, two cities that lie close to each other are drawn on a small-scale map as one city. This is the most radical type of generalization because it involves the creation of a new object, with connections to the contributing objects. Nevertheless, there is no reason why it should not be possible. The resulting GIS will become more complex.

9.6   Discussion

The design of the presentation of census data studied in this chapter is based on the principle of data abstraction. Inheritance, a powerful feature in most OO programming languages, is not used here. However, there is at least one situation in which it could be very useful. In Section 9.2, five object types were introduced of which only one, the economic geographic region, was described. The others, province, corop region, nodal region, and municipality, can be described similarly. That is, large parts of the object type are exactly the same. This can be handled by an abstract object type (generalizing class) called Administrative Unit. The actual object types inherit large parts of this abstract object type and only the specific parts are local. This approach would also be better for the maintenance of the system.

Of course, the OO approach does not solve every problem that is encountered during the design of a system. For example, the use of the R-tree and the BLG-tree is necessary to meet the requirements of a reactive information system. The properties of these data structures are also valid outside the OO approach. Nevertheless, the OO approach offers a good design environment that is also well suited for GISs. In order to test the quality of the design, the presentation of census data will be implemented as a case study. General-purpose GIS object libraries will be built and used in the course of this study. This will reveal more about the applicability of the OO approach to the whole development process of a GIS.

The disadvantage of the presented solution is that a file or DBMS is still required for the long-term storage of the contents of the objects. This problem can be solved by introducing persistent objects in Procol, as described in the next chapter.

10 Persistent Procol

Persistent objects are objects whose contents may outlive the execution time of the program. This chapter describes the process of introducing persistent objects into the object-oriented programming language Procol (see Section 9.1). The strength of persistent objects in an object-oriented programming language is the integration of a database system with a programming language. Persistent objects make the program development easier, because the programmer does not have to implement the explicit loading and saving of data. Besides the general functional aspects, special attention is paid to graphic applications in order to deal with their specific geometric requirements. For example, it must be possible to find, in an efficient manner, all graphic objects that fall within a given region. These issues, persistent objects and their geometric requirements, have not yet got the attention they deserve in the literature covering object-oriented graphic systems where modelling and functional aspects dominate.

The fact that the object-oriented approach is so successful in computer graphics is due mainly to the system modelling capabilities that the object-oriented paradigm offers. The specialization relationships that exist between the graphic objects in a system can be modelled with inheritance or delegation, one or both of which are present in many object-oriented development environments. Geometric data types, such as points, vectors, and matrices, may be implemented by abstract data types using object classes (Blake and Cook 1987, Cox 1986, Dietrich, Jr *et al.* 1989, Meyer 1988). However, in practice this modelling power is not enough when implementing CAD systems or GISs that deal with large data sets. Integrated database facilities are required to support the graphic application in an efficient manner.

Our solution is based on the introduction of persistent objects in Procol. The resulting system is extendable with new (persistent) types and operators. In itself, this approach is not new and has been described by several other authors (Atkinson and Buneman 1987, Richardson and Carey 1989, Straw *et al.* 1989), but we also tried to make the system suitable for highly interactive and graphic applications. This goal is achieved by putting emphasis on both time efficiency (offering techniques such as navigation, index structures, and parallelism) and the recognition of multi-dimensional data. As an example, not only one-dimensional, but also multi-dimensional index structures are provided in Procol.

Section 10.1 states the problem area and Section 10.2 defines our general requirements for persistence in an object-oriented language. Our first attempt to introduce persistence in Procol and comparisons with several other systems are described in Section 10.3. Problems concerning referential integrity are described in Section 10.4. Building on this experience, the following topics are discussed in more detail in Section 10.5: object identity, (multi-dimensional) search problems, programming interface, and object instances of different sizes. Section 10.6 presents the solution as chosen in Procol. In nearly all the sections, the requirements of graphics play an important role. The discussions are illustrated with examples based on graphic systems.

10.1 The Need for Persistence

The computer science research in the area of programming languages emphasizes programming constructs and data structures. One of the most popular paradigms is that of abstract data types (ADTs). Object-oriented programming languages (OOPLs) encapsulate this paradigm in an elegant manner using object types to describe the ADTs. Access from outside to the data inside an object instance is possible only through the methods or procedures defined for that object type. Note that in this chapter I will use the term 'object type' to indicate a class of objects and the term 'object' to indicate one instance. In case more emphasis is needed I use the term 'object instance' explicitly.

The data stored in data structures (or objects) of a running program are in general volatile; that is, as soon as the program stops, the data are lost. However, in many applications the data themselves are very important. An obvious solution is to save the data in a file by explicit write statements. The next time the program is started it first reads the data from the file into the volatile data structures. Persistent objects make the program development more efficient, because the programmer does not have to worry about the reading and writing of data from and to disk. Also, the structure of the data file may become quite complex, possibly resulting in intricate parsing. In the context of CAD systems, Cockshott (1987) claims that 'a third of all programming time is expended ensuring that vital data is safely copied to disk and retrieved when needed'. Moreover, for large quantities of data this 'file' solution becomes cumbersome during the execution of the program. Consider as an example an information system that registers bank accounts. A characteristic of this and many other information systems is that the objects are well structured, quite passive, and occur in large quantities. Passive objects are objects that hardly ever send messages to other objects (except for replies); they only react to messages from outside. In the bank account example, an account object replies its current amount when asked, and updates it when told to do so by a message from an authorized object.

Database management systems (DBMSs) have been developed to deal with the large amounts of data mentioned above. DBMSs concentrate on the information representation and tackle related problems such as integrity, security, redundancy, consistency, efficient searching, query formulation, and concurrency control. A major drawback is that the DBMSs of the current generation are not extendable with new data types and operators. This makes the use of these DBMSs inconvenient in non-standard applications that need support for other data types.

The database research community has recognized this deficiency and is now trying to design systems that are more open (Egenhofer and Frank 1989a, Stonebraker and Rowe 1986, Wolf 1989). At the other side, the OOPL research community has recognized the need for persistent objects. Now these two worlds meet. In some cases these encounters result in conflicts because of very different and incompatible principles. For example, explicit (navigation) links among instances are considered harmful by the database community, because they are hard to maintain. However, these same links form the backbone in representing complex objects in OOPLs. On the other hand, sometimes the combination of the two research communities results in a nice symbiosis. An example of this is the concurrency control problem in the database, which is solved by the model of an object that accepts messages one by one (as in Procol).

We are interested in interactive graphic applications, such as CAD systems, VLSI Design, and GISs. Common aspects in these systems are: interactivity, graphics, and large amounts of data. In the previous chapter it was seen that the object-oriented approach offers a good data modelling and design environment for GIS applications. Also, the need for persistent objects in our object-oriented programming language Procol was identified.

10.2   A Wish List

This section briefly considers the major requirements for persistent objects in Procol. Some will be elaborated on in one of the later sections, in which case the section referred to will be mentioned here. Note that these requirements may be neither orthogonal nor independent of each other.

pp1 Upward compatible. Persistent objects have to be introduced with a minimal change to Procol. This implies that existing Procol programs do not have to be changed in order to be compiled by the new version of the Procol compiler.

pp2 Transparent persistent objects. Persistent objects are treated in the same manner as volatile objects by the application, except perhaps for incompatible database facilities, for example associative searching, that is object searching based on the contents or value of instances. Atkinson *et al.* (1983) refine this by recognizing the following principles for persistent data (they assume several levels of persistence):

1. persistence independence: the persistence of an object is independent of how the program manipulates that object; so it has to be possible to call a procedure of which the actual parameters are sometimes persistent objects and other times volatile objects;

2. persistence data type orthogonality: all objects are allowed the full range of persistence; this means that, no matter how complicated

the type is, its instances can still become persistent.

pp3 Complex objects. This provides the system with sufficient modelling power, e.g. part-of hierarchies. In Procol complex object types are defined by means of links (or references) to other object types that together define the complex object. The complex objects are static in terms of the object type structure and dynamic in the sense of the object instances.

pp4 Extensibility with new ADTs. This wish might be a trivial one in the context of OOPLs or object-oriented databases, but certainly not in the context of the traditional DBMSs. The definitions of the new persistent ADTs also have to be stored somewhere, if we want to be able to manipulate their object instances in a sensible manner.

pp5 Efficient handling of large amounts of objects. Long-lived systems allow time for data to accumulate. This, combined with the fact that we aim at developing interactive systems, makes this efficiency requirement even more important than in other systems. Not only is efficient retrieval by object id (which is very important in OOPL and object-oriented databases, as used in navigation links) required, but also efficient associative searching has to be possible. This is realised, as usual, by indexing techniques such as B-trees (Bayer and McCreight 1973, Comer 1979) or hashing.

pp6 Object instances of different sizes. A polyline or polygon has to be stored with a minimum of overhead, because of the required time (and space) efficiency in interactive systems. This implies that different instances of the same object type may have different sizes. To treat an object instance as a unity means that it is stored in a contiguous part of memory. This may seem to be an implementation issue, but it is very important, and by putting it on our wish list we emphasize this. This topic is further discussed in Subsection 10.5.5.

pp7 Highly interactive and graphic applications. The previous two wishes are actually part of this more general wish to make Procol suitable for this kind of application. It has to be kept in mind that multi-dimensional data sometimes require other approaches than the data types encountered in traditional DBMSs. Also, the fact that Procol is designed as a parallel programming language should be exploited.

pp8 Exchangeable objects. It should be possible to exchange object instances between different systems. Object instances created by one system must be directly applicable by other systems.

pp9 Ability to deal with referential integrity in a satisfactory manner. This is a well-known problem in database and programming language research. The topic will be discussed in depth in Section 10.4.

10.3   The First Attempt

In this section I shall briefly compare our first (and quite simple) attempt, the 'snapshot' method (see Section 9.3), with the approaches taken in some other systems. Note that the 'snapshot' method to provide a mechanism for persistent objects does not try to satisfy all the wishes of our list. The problems concerning object management, encountered in Section 9.3, give some indication of the complexity of introducing persistent objects.

Methods similar to the 'snapshot' method are used in several other OOPLs. In systems offering multiple inheritance, an object type, which also has to be persistent, inherits this property from a general object type with methods to save and load the object. Egenhofer and Frank (1989*b*, 1988*a*) suggest the object type `db_persistent` with methods `store`, `delete`, `retrieve`, and `modify`. ET++ (Weinand *et al.* 1989) has an object hierarchy with the object type `object` in the top of this hierarchy. The object type `object` has methods called `PrintOn` and `ReadFrom` which enable transfer to and from disk. These solutions work well as long as the object types contain no references to other objects but only simple attributes, such as an array of coordinates describing a polygon.

The persistent data in PS-algol (Atkinson *et al.* 1983, Morrison *et al.* 1986) are organized into one or more databases. Each database has its own root and may contain values of different (complex) data types. The data are 'imported' into a program with the `open.database` procedure which returns a pointer to the root. The root has the form of a name-value table in which the value is usually a pointer to another data structure. The actual data are accessed by following these pointers, and it is assumed that the programmer has to know the structure of the database (though this is nowhere stated in the PS-algol papers). Once imported, the data can be manipulated in the same manner as volatile data. The procedure `commit` propagates the changes made thus far to the database, if it is open for writing. Everything that is accessible from the root is stored. This means that values may change and data (structures) may be added or removed.

In the OOPL Eiffel (Meyer 1988), an object type that inherits from the object type `STORABLE` acquires this kind of behaviour by means of the methods `store` and `retrieve`. If the method `store` is invoked in object instance `x`, the whole object structure starting at `x` is dumped (in a special format) in a file, even if the referenced object types in `x` do not inherit from `STORABLE`. Depending on `x` and the object structure of the application, it is possible to store the whole object structure, or just a part of it. Basically, this solution has two drawbacks. First, the application programmer has to indicate when to save or load the objects explicitly. So, if the program is stopped before the 'save', the latest data are lost. Second, updating one object in an object structure can become very expensive if all related objects have to be saved also, even if they have not changed.

10.4   Referential Integrity

What happens if an object is deleted by its creator while other objects are still referring to this deleted object? A dangling reference is not a problem specific to persistent objects; it is a problem in the case of volatile objects too, but it manifests itself in a severe manner in combination with persistent objects. Assume that a persistent object contains a reference to a volatile object and the program is stopped. The next time the program is started, the reference to the volatile object is not valid any more (though it has not been deleted). Dangling references can also occur in non-OOPL. For example, in C it is possible to have pointers to deleted data structures, which may be the cause of some severe errors in a program. Some systems guarantee referential integrity. An associated problem is that of an unreachable object, that is an object to which the last reference is lost. There are a number of possible approaches towards these problems:

1. If we want to guarantee referential integrity, we have to be able at least to detect whether the integrity is damaged by the deletion of an object. This can be achieved by associating a reference count with each object instance. If the reference count has a value greater than zero, the object will not be deleted and the creator is notified of this fact. The reference count mechanism introduces overhead, because the counters have to be updated in each assignment to an object variable. Problems are introduced by cyclic data structures.

2. A slightly different approach, but also based on a reference count, is followed in O_2 (Bancilhon *et al.* 1988). The object deletion is not refused but is postponed until the value of the reference count is zero. The creator does not have to worry about trying to delete the object another time.

3. Dangling references cannot occur if we prohibit the deletion of objects. This approach is taken in for example GemStone (Penney and Stein 1987). In order to avoid congestion of the system, garbage collection has to be carried out. Two well-known methods for this are:

- using a reference count: when the count becomes zero, the associated object instance is automatically deleted by the system;
- performing a sweep through the object space (a directed graph) in order to detect which objects are unreachable. A disadvantage here is that the sweep is performed periodically and during this operation the system cannot be used by the applications. This can be avoided by using an incremental version of the sweep algorithm.

4. The maintenance of the reference count introduces overhead; both memory usage and execution time increase. Clearly, it is more efficient to omit a reference count and directly delete the object at request. However, in this case the system is not allowed to reuse the ids of deleted objects for new objects. So, if a message is being sent to a deleted object, the system can detect this, and the sender will be notified. This strategy implies that the address of an object cannot be used as its id, because when the object is deleted we want to be able to reuse that part of the memory space for new objects.

It may be clear by now that we are biased towards the latter approach. In the context of Procol, dangling references are probably programming errors and the detection of the illegal use of dangling references during runtime is an adequate solution. Finally, it is interesting to note that PCTE+ (IEPG TA-13 1988, 1989) offers links both with and without referential integrity. This is probably done for efficiency reasons. It is not stated in the PCTE+ documents how the referential integrity is maintained.

10.5   The Object Management System

This section presents some issues concerning the semantics of constructs which have to be added to Procol to support persistent objects. This is done here without worrying about the exact syntax or how this can be achieved in our implementation of Procol. In order to solve the adminis- trative problems associated with the use of object ids, there is a need of an object management system (OMS) that takes care of the (persistent) objects. One of the responsibilities of the OMS is to keep the references in the object system consistent. To be more precise, an object system is consistent if (Khoshafian and Copeland 1986):

- no two distinct objects have the same identifier (unique identifier as- sumption); in other words, the identifier functionally determines the type and the value of the object;
- for each identifier present in the system there is an object with this identifier (no dangling identifier assumption).

10.5.1   Object Identity

A uniform object identification mechanism has to be developed which is capable of dealing with objects shared by multiple programs, multiple users, or even multiple computers (in a network). There should be a mechanism to indicate in which persistent objects one is interested, so that one is not bothered by non-interesting objects of others. One possible method could be to organize the object instances in 'data sets' which are put in the normal hierarchical file system. This limits the scope and makes the task of finding the right communication partner easier for the OMS.

In relational databases (Codd 1970) an identifier key is formed by one or more user-supplied attributes. Value-based matching is a transparent technique for expressing relationships. However, it provides no support at all for referential integrity. By contrast, OOPLs support the notion of object identity which is independent of the attribute values (Paton and Gray 1988). Khoshafian and Copeland (1986) describe several techniques for implementing object identity and they conclude that using so-called surrogates is the best technique. Surrogates are system-generated, globally unique identifiers, completely independent of the physical location and data contents of an object.

10.5.2   Searching

The objects presented so far are not suited for associative search operations, that is, for searching based on the contents of an object instead of using the object id to find an object. This is especially useful for a program that wants to use objects created by other programs, because the ids are unknown and have no semantic meaning. All that a program(mer) knows about is object types (the kind of data he wants to use) and attribute values (restriction of instances).

Another use of associative searching is for solving efficiently alpha-numeric queries, such as (see Section 9.3): 'How many inhabitants are there in the municipality with the name Leiden?' Section 9.3 suggested using the B-tree (Bayer and McCreight 1973, Comer 1979) for this purpose. The B-tree solution in an object-oriented environment is established by a set of auxiliary (system) objects. These objects do not contain the application data, but they do contain tree structures with references to the objects with the actual data. This B-tree has to be part of the OMS and, if possible, transparent to the 'application' objects. Note that the OMS itself can be implemented in Procol as a set of objects.

There is some friction between the concepts behind the ADTs and the idea of associative searching, because associative searching requires access to the internals of other objects. An object has to specify a query in terms of the data-part inside another object. To limit the damage, only so-called visible attributes may be used in the query. These visible attributes become part of the specification of an object type (together with the actions, of course), in contrast to the non-visible data-part which belongs to the implementation. Note that an index may be put only on a visible attribute of an object type.

10.5.3 Multi-Dimensional Data

Section 9.3 stated that the searching problem also applied to the graphic or geometric data. Not all known spatial data structures (van Oosterom 1988*a*, 1988*b*) are suited for this purpose. For example, KD-trees (Bentley 1975), Quadtrees (Samet 1984), R^+-trees (Faloutsos *et al.* 1987), BSP-trees (van Oosterom 1989), Cell trees (Günther 1988), and Grid Files (Nievergelt *et al.* 1984) are more difficult to integrate in the object-oriented environment because they cut the geographic objects into pieces. This is against the spirit of the object-oriented approach, which tries to make complete 'units', with meaning to the user. The R-tree (Faloutsos *et al.* 1987, Greene 1989, Guttman 1984), Field-tree (Frank and Barrera 1989), Reactive-tree (van Oosterom 1991), KD2B-tree, and Sphere-tree (van Oosterom and Claassen 1990) are good candidates for integration in an object-oriented system, because they do not split the objects. Each of the spatial search structures has its own strengths and weaknesses, so if several alternatives are offered by the system, an application can use the structure that fulfils its needs the best.

In Procol, trees can be implemented in two different ways. The first method stores the entire tree in one single search object. The second method stores each node of the tree in a separate instance of the search object. The latter introduces overhead by creating a lot of search objects (nodes). However, it has the advantage of being suited for parallel processing in Procol because the search objects can run on parallel processors.

This is useful for range queries such as: 'Give all municipalities that fall within a specified rectangle'. Appendix A contains a part of the Procol implementation of the R-tree. In this implementation each node is a separate object. In the case of the R-tree this is a reasonable choice, because there is a fair amount of work in each node. The same would be valid for B-trees, but not for binary trees (see Section 9.3). In any case, for practical reasons, there has to be a separate search tree (index) for each attribute for which efficient searches are required. This has to be made clear to the OMS before the queries are posed.

10.5.4   The Programming Interface

From the user's point of view, the extension should be as small and simple as possible. First, we have to decide how to indicate that an object is persistent. Some alternative possibilities are:

- After creation: Make a volatile object instance persistent by applying a new Procol primitive **persistent**. Assume the variable x holds the id of an object instance; then this instance is made persistent by: **persistent x**. Note that this is different from the 'snapshot' method of Section 9.3, because the values (states) of a persistent object are always guaranteed to be up to date.
- Per object instance: At the moment an object is created, it is decided whether or not it will be persistent. A convention can be made that objects created with the **new** primitive are volatile and the ones created with the **persistent** primitive are persistent.
- Per object type: At the moment the object type is defined, it is specified whether or not all instances of this type are persistent. A modified keyword **OBJ** could indicate this: **PERSISTENT_OBJ**.

A combination of these approaches is also possible. In the language E (Richardson and Carey 1989) the programmer has to indicate per type (class) that instances are optionally persistent; the programmer has to decide per instance if it is really a persistent object instance.

The advantage of persistence per object type is that only once, during the object type definition, is there a difference for the application programmer between persistent and volatile objects. In the other solutions it is required to indicate that the object is persistent for each object instance. The major drawback of persistence per object type is that two different types have to be defined if we want to use both the volatile and the persistent variants of basically one object type. In the case of strong type checking this means that we cannot freely interchange the use of volatile and persistent objects as arguments in messages and procedure calls.

In order to get hold of persistent objects with unknown id, Procol will be extended with the **retrieve** primitive. Perhaps it is better to take the following approach towards the primitives **new, delete,** and **retrieve**: consider them as messages to the object types themselves ('class methods'). These are system objects that are (partly) responsible for the OMS tasks.

These system objects have to maintain index structures if requested by sending them an **index** message.

The **retrieve** primitive has some resemblance to the **new** primitive, because it also assigns the id of an object to a variable of the proper type. Unlike **new, retrieve** will not execute the **Init** section, because that already happened when this object was created for the first time. The protocol (expression) of the object regulating access to the object is matched starting at the current (saved) state. A discrimination condition can be used, because the object type information may not be specific enough. Of course, only visible attributes can be specified in the condition. A retrieve returns the id of the object of the proper type for which the discrimination condition evaluates **true**. If there is more than one object satisfying these criteria, only one is returned; the other objects can be obtained by using the **next** operation. If there is no object satisfying these criteria, a **NULL** object is returned.

It is a small step from the **retrieve** primitive to the associative search operation. In fact, it could be considered as an iteration over the retrieve operation. If fast replies are required, an index has to be used in the case of a large set of objects. This index could be a spatial index structure, e.g. needed for efficiently solving a 'rectangle' query. There are several options for returning the answer of a search:

- Return one big set that contains the ids of all objects that satisfy the query. In the case of a large number of answers, a lot of temporary memory is required and it may take quite a while to generate the complete answer.
- State the query first and then retrieve the answers one by one (or perhaps in buffers of a fixed size). The first part of the answer will probably be ready sooner than the complete answer would be. This promotes parallelism and is also quite important in an interactive application, because the end-user could already be presented with a partial answer on his screen.

The problem with the second solution for returning a search result is that other objects might interfere with the set of objects that belongs to the queried object type. 'Third-party' objects could change values and add new instances or delete existing ones. This can be solved by applying the proper protocols in the system (OMS) objects.

10.5.5   Object Instances of Different Sizes

In Section 10.2 we saw that the wish to store a polyline or polygon with a minimum of overhead implied that different instances of the same object type may have different sizes. So, for example, the pure relational solution presented by Van Roessel (1987) is not acceptable, because a polyline is scattered over several tuples in a table and has to be aggregated before it can be used again. In Chapter 7 (van Oosterom *et al.* 1989) a solution in the context of the relational data model has been presented.

Different sizes have an (enormous) impact on the implementation of persistent objects. In CO_2, the C implementation of O_2 (Bancilhon *et al.* 1988), it was decided to prohibit object instances of different sizes. In contrast, we would even like to have persistent objects whose sizes change dynamically, for example to make it possible to remove points from or add points to a polyline. However, this would complicate the implementation even further. A decrease in the size of an object is not too hard, but an increase in object size means that an object does not fit into the (contiguous) part of its memory and the memory after this object is probably occupied by another instance. The object will have to be moved to a larger place, because we want to treat it as a unity and do not want to split it. This would have been impossible if the object id were its address. However, this was already disapproved for reasons discussed earlier. In any case, growing persistent objects could introduce a lot of overhead, e.g. moving of objects.

A more feasible situation is that after the **Init** section the size of an object may not vary any more. It is still possible to deal with dynamic problems; for example, use a pointer (id) to an object of type linked list. This object type has an 'application' data-part and a pointer to the next list element. Each instance represents one list element and they all have the same fixed size. We can extend this approach and simplify our implementation of Procol, if we only allow the following data types as attributes: basic types (int, char, float ...), references (or links) to other objects, and arrays with fixed size after the **Init**.

10.6   Persistent Objects in Procol

The question how to implement persistent objects in Procol can be divided into two sub-questions. The first is how to adapt the language features (syntax: the external implementation). The second is how to implement this on the underlying platform (the internal implementation).

10.6.1   External Implementation

Our decisions concerning the external implementation of persistence in Procol included the introduction of the following new keywords:

1. **persistent** <object-id>
   in <dataset-key>
   With this statement a volatile object instance identified by <object-id> can be made persistent by coupling it to a data set identified by <dataset-key>.

2. **volatile** <object-id>
   With this statement an object instance (identified by <object-id>) that has been made persistent before can be made volatile. The result of this statement is that the persistent object instance is removed from its data set.

3. **retrieve** <object-id>
   from <dataset-key>

where <discrimination-string>
and
next <object-id>
With this statement we can retrieve an instance of the same type of
<object-id> from the data set with identifier <dataset-key> that sat-
isfies the <discrimination-string>. If the data set in question contains
more than one instance of the required type, only one will be returned
in <object-id>. Any successive execution of the **next** statement will
retrieve the next instance of the required type until all required instances have been retrieved from the data set.

In general, <dataset-key> will be a string. This string can then be used
to compose the file names of the necessary data set files. The following is
an example of the provided persistence in the **Declare** section of an object
type using the object DRAWING:

```
Declare          object DRAWING drawing;

                 allocate_drawing(char *name)
                 {
                     new drawing;
                     persistent drawing in name;
                 }

                 read_old_drawing(char *name)
                 {
                     retrieve drawing from name;
                 }
```

10.6.2 Internal Implementation

We will now motivate our design decisions for the internal implementation.
Procol has been developed and implemented on a network of Sun workstations running under SunOS Release 4. Five possible implementations of
the extension of Procol with persistent objects were considered (compare
with (Tsichritzis and Nierstrasz 1988)):

1. Bare implementation: The problem of persistent objects is deferred
to the kernel of Procol. Procol stores and manages these objects in (one
or more) files, by using the Unix OS-interface calls: **open**, **close**, **read**,
and **write**. The advantage of this method is that it is very portable.
Disadvantages are: a lot of work, a major increase in size of the Procol
compiler, and relative inefficiency because of the many **read/write** calls
that have to be made. We also have to implement the memory management aspects (within each file), such as allocation, deallocation, gaps, fit
strategies, and so on. 2. Shared mapped memory (virtual files) implementation: A file is mapped directly by the Unix OS on the address space
of a process. A major advantage is that there is no difference between
the 'stored' object instances and their 'running' counterpart, at least not
at the level of the Procol kernel. At OS level there is a difference, and

this is the same as the difference between virtual memory pages that are in main-memory and the ones that are swapped on disk. We expect this implementation to be very efficient. In order to gain some experience, we are currently converting a local application that uses explicit **read** and **write** statements into a mapped memory implementation. Test results indicate that the elapsed times decrease by about 30 per cent in applications with a lot of **read** and **write** statements. A disadvantage of the mapped memory approach is that we still have to do the memory management ourselves. 3. Ingres (Sun Microsystems, Inc. 1987) (or other relational DBMS)-based implementation: The memory management is now done by database system. Also, all kinds of other useful DBMS features are provided. Wiederhold (1986) suggests the use of view-object generators to reconstruct objects from relations and view-object decomposers/archivers for the inverse operation. However, there are several disadvantages. Object types cannot always be mapped on a relation in which each tuple represents one object instance; for example, objects with polylines have changing instance sizes. Another disadvantage is that there is a difference between the 'stored' object instance and its 'running' counterpart, as observed by the Procol kernel; — although not, of course, from the point of view of the application program. 4. Postgres (Rowe and Stonebraker 1987, Stonebraker and Rowe 1986, Stonebraker *et al.* 1990, Postgres Research Group 1991) (or other extendable DBMS)-based implementation: New data types can be defined, so it should be possible to have a one-to-one relationship between tuples in a relation and the objects. Still, the difference stored/running parts of objects remains. Another disadvantage is that a DBMS implementation is probably less efficient than the mapped memory implementation. 5. PCTE+ (IEPG TA-13 1988, 1989)-based implementation: The Portable Common Tool Environment PCTE+ offers a powerful OMS derived from the entity-relationship model of Chen (1976). The objects in PCTE+ are not complete encapsulations of ADTs. An advantage is that the OMS of PCTE+ is already distributed among workstations in a network. It has to be seen if it is possible to manipulate multi-dimensional data in an efficient manner. A disadvantage of a PCTE+ based implementation is that we have to switch completely to the PCTE+ concepts and will be tied to these concepts. Finally, the availability of PCTE+ may be a problem. It is not in the public domain, and therefore we expect problems with the distribution of Procol.

We have decided to use the mapped memory approach for the prototype implementation of persistent objects in Procol. The main reasons were the expected efficiency of the approach and the provided synchrony between 'stored' and 'running' object instances.

An object instance is identified by a surrogate (Khoshafian and Copeland 1986). That is, the object id is not the actual address in memory, but we use an indirection (Straw *et al.* 1989) to locate the object; see Fig. 10.1 for a graphic demonstration of the process. Each surrogate contains an indication of whether the object instance is volatile, persistent, or deleted.

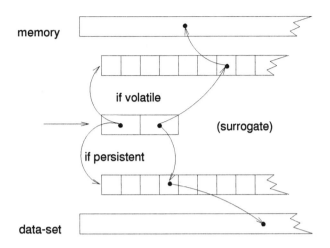

FIG. 10.1. Object Reference with Surrogates

When, during the execution of a Procol program, an object is referenced, we have to check the first part of the surrogate. If the the object instance is volatile, the surrogate contains a key that can be used to retrieve the memory address of the instance variables. If the object instance is persistent, the surrogate contains a data set identifier and a key. With this data set identifier and key, the OMS is able to retrieve the actual memory address of the the instance variables of the persistent object in question. The disadvantage of surrogates is that there is an extra indirection. However, surrogates have several important advantages. They enable objects to be switched between volatile and persistent states, by modifying a part of the surrogate. We decided, on the basis of advantages and disadvantages mentioned in Section 10.4, that the control of referential integrity is not required in Procol. However, the use of illegal references is signalled at runtime, by inspecting the surrogate and taking the appropriate measures.

If a persistent object contains a reference to a volatile object, this volatile object is not saved automatically. The proper way to program this case is to make the other object also persistent, if the referenced object is still needed in the future. The state of each object instance is stored as unity, that is, in contiguous memory. Instances of different sizes are no problem. The state also contains some additional data, for example the **Creator**.

The application programmer decides whether instances of the same object type are 'stored together' without instances of other types in one data set, or if a data set contains instances of different types. The latter is advantageous for representing complex objects in an efficient manner, because the instances of the different types that define a complex object are stored close together. Note that complex objects are important in CAD systems, because this is one of the main modelling tools. Associated with each data set of instances is a header which contains, depending on the implementation method, some additional information. For example, in the case of surrogates, the header contains the address lookup table based on the object id (in the form of a B-tree). The header is also the place where index structures are stored and where the memory management information is kept. Summarizing, the major differences between a table in relational DBMS and a data set are:

- A table contains only tuples of the same type, in contrast to a data set which may contain instances of different types.
- There is an explicit object id in our system instead of a primary key, which is an attribute (or combination of several attributes).
- Attributes may be references to other objects. This is allowed for both object instances of the same object type and for instances of another object type.
- The instances are not necessarily of the same size, though the number of attributes is equal.
- The way in which the tables are manipulated is different. This is not limited to a language such as SQL, but general program language constructs can be used (complemented with a number of specialized operations).

10.7   The Case Study

In this section the presentation of census data, as discussed in Subsection 1.5.2 and implemented in Procol as described in Chapter 9, is modified into a Persistent Procol application. As stated in previous section, the only needed changes in code should be:

- To create a persistent object, the statement
  persistent <objectname> in <dataset>
  has to be inserted after the statement:
  new <objectname>.
- Instead of creating the whole hierarchical administrative unit structure, the statement
  retrieve country from <dataset>
  has to be added.

The first step taken in the process of modifying the Procol program into a persistent one was to split up the GIS and the data reading part into two separate programs. The final GIS program contains no **read** or **write** statements: it only refers to persistent objects by retrieving them from a data set. As an aside, by splitting the program into two separate applications it was demonstrated that persistent objects can be shared and transported between different applications. Fig. 10.2 shows the design of the resulting programs.

The data translation program creates a data set for every object type in the map structure; this shows that references to persistent objects can extend to data sets other than the one where the reference actually takes place.

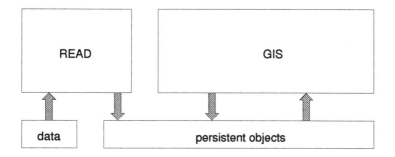

FIG. 10.2. Design of the Example GIS

The resulting persistent data structure has the same hierarchical structure as the data structure in the non-persistent version. It has one single root, the country. The country has references to all existing provinces, and these provinces contain references to the objects at the next lower level (corop regions), and so on until the municipalities are reached. This type of data structure can be identified by one single reference root. Retrieval of the data structure is simple, as one only has to get hold of the root of the structure, in this case the COUNTRY object.

It is not a prerequisite of data structures that they should be hierarchical to be used in this persistent manner. The data structure itself may be cyclic, and the data structure may well take the form of a graph. It is important that we are able to identify a certain entrance to the data structure. In this example, the entrance is formed by the COUNTRY object. As it is the only instance of its type, it is possible to retrieve it from the "holland.pp" data set uniquely. It would have been possible also to define an extra entrance in the form of, say, an index structure on the name of the municipalities. This would have enabled the GIS to make selections based on municipality name and to highlight the corresponding municipality on the map, for instance. This index structure could have been retrieved from a data set based on its type, in the same manner as we retrieved the country. In this case we would have had two entrances to the same data structure, and different interpretations and usage of the data structure could be envisioned.

10.7.1 Retrieving Persistent Objects from a Data Set

The data structure is retrieved from its persistent storage by retrieving the COUNTRY object instance (the root of the data structure) from its corresponding data set (i.e. "holland.pp"). The assumption is that the data set contains only one instance of the type COUNTRY, and issuing the retrieve primitive will produce the one and only COUNTRY object instance forming the root of our data structure. If the data set contains more than one instance, one among them is returned. If desired, the next primitive can be used to retrieve the other COUNTRY objects from the data set. Retrieving the country is the only thing that has to be done in order to retrieve the data structure from its persistent storage. That is because the `COUNTRY` object contains references to a number of persistent objects. While accessing these references, for instance while drawing the map, the Procol runtime system will detect that these references are references to persistent objects. The corresponding data sets are then mapped to main memory (by the operating system) and the persistent objects are automatically loaded.

Only two lines of code need to be changed in the object that initiates the implementaion of the case. The object is called the `GIS` object. The non-persistent version of the `GIS` object reads in the country by creating a special `READ` object instance, shown in the following code segment:

```
new read;
request read.GetCountry() -> (country);
```

The persistent version of the case retrieves the `COUNTRY` object instance from a data set, and that is all that needs to be done to make the GIS persistent:

```
retrieve country from "holland.pp";
```

Of course, the objects created by the `READ` object need to be made persistent and stored in a data set before they can be retrieved in the manner shown above. To begin with, the `READ` object will not only create the `COUNTRY` object instance, but will also make it persistent. This is done in the **Init** section of the following segment:

```
Obj     READ
Declare ....

Protocol ....
Init    new country; persistent country in "holland.pp";
        ....
Actions ....
Endobj  READ;
```

Furthermore, the other read objects, ranging from province to municipality read objects, also have to make persistent the objects they create. Summarizing, to make the non-persistent GIS persistent, one `retrieve` statement and few `persistent` statements suffice.

10.7.2   Persistent Program Efficiency

The translation program and the GIS application have been implemented and timed with `/bin/time` on a Sun Sparcstation IPC, with SunOS Release 4.1.1 and Sun OpenWindows. In the non-persistent version it takes 40.1 seconds to start the program, display the map, and finish the program again. The persistent version takes only 4.7 seconds for these steps. The difference is introduced mainly by the time needed to convert the relatively unstructured data into a structured set of objects. To be honest, converting these into persistent objects takes even more time in the current implementation, as the process of translating volatile objects into persistent objects is not yet fully optimized. However, this is the once-only time needed to build the database. After building it (i.e. the co-operating set of persistent objects), the database can be used in a very efficient manner.

It is significant that applications based upon persistent objects do not contain **read** or **write** statements to load or save objects. Furthermore, the data file that serves as input to the application need not be parsed. This is because the persistent objects contain references to each other and the inherent structure in the data file has been determined before and translated into persistent object references.

Although speed increases when introducing persistent objects, the size of the data sets, as a drawback, is increasing too. This interchange of speed and size is not specific to persistent objects: it is, for example, a general problem for compilers; code is optimized for speed or size. In our case, the size of the map data in binary format is approximately four times smaller than the size of the data sets containing the persistent objects of our map data structure. This can become a problem when dealing with very large quantities of digital geographic data.

10.8   Discussion

Besides an object-oriented programming language, Procol is also a parallel programming language. It is possible that the objects run in parallel on multiple processors. I have used this only in a few examples; see Appendix A. Clearly, this topic deserves more attention, and more research is needed in the context of highly interactive and graphic systems.

It is a small step from one single-user Procol program with persistent objects to a system with multiple users. At least conceptually, because each object has its own protocol which regulates the communication, it should not matter from which program a message originates. However, we will have to reconsider some of the concepts.

The provided query facilities are very limited; with **retrieve**, it is only possible to get hold of one 'starting' object id of a specified type or to perform the selection of instances from one set (object type) at a time. More complex queries have to be programmed into the objects (the Procol program). Care has to be taken to avoid object types becoming too specific. This would be in contradiction with one of the basic principles of databases being independent of applications. More research in this area is necessary.

Conclusion

Summary of Important Results

The explicit formulation of the six requirements of GIS towards the data model (*r1* integrated storage of geometric, topological, and thematic data, *r2* spatial capabilities, *r3* levels of detail, *r4* layers of entities, *r5* persistence, and *r6* dynamic) and the categorization of spatial data structures (into geometric data structures, topological data structures, and data structures with detail levels), as given in the first two chapters of this book, form the first preliminary research results. In the Chapters 3–6 several new spatial data structures were presented, each having their strengths and weaknesses. In Chapter 3 a modified BSP-tree for GIS was suggested and implemented. This implementation showed that in the GIS practice the BSP-tree requires only linear storage instead of the worst-case quadratic storage. Further, this chapter presented the first reactive structure: the reactive BSP-tree, a static structure.

Two new geometric data structures, the KD2B-tree and the Sphere-tree, that are orientation-insensitive, were introduced in Chapter 4. Both structures use less storage space than the well-known R-tree, and an implementation showed that the KD2B-tree in particular performed very well: the R-tree was outperformed in case of sphere and polygonal overlap queries, and the KD2B-tree was very competitive in case of rectangular queries. Chapter 5 presented a new data structure with detail levels, the BLG-tree, which has some clear advantages over known structures, such as the Strip tree and the Arc tree. One of the most important research results of this book is the Reactive-tree, a fully dynamic and reactive data structure described in the last chapter of Part I. The future will reveal whether the Reactive-tree will become a storage structure for a scaleless, seamless geographic database.

Part II described the research related to geographic data modelling and persistence. Chapter 7 demonstrated that a GIS based on the relational model is possible, is as far as it concerns most of the functional aspects, but not from an efficiency point of view. Therefore, a geographic extension was proposed and implemented in Chapter 8. Then in Chapter 9 Procol was used to show that the OO approach offers a better modelling and development environment. It was explicitly indicated how typical GIS issues may be addressed (geometric searching, map generalization, topology, complex objects) as illustrated by the example of census data presentation. The most important research result of Part II was the introduction of persistent objects in Procol in Chapter 10, accompanied by the description of a new implementation technique based on shared mapped memory. The inclusion of index structures such as the Reactive-tree leads to a powerful development environment and a basis for GIS applications capable of dealing with instances of different sizes, efficient geometric searching, large data sets, complex objects, referential integrity, exchange of objects between two systems, and so on.

Future Research

In the discussion sections of the Chapters 3–10, several suggestions are made for future research. In this section, the most important ones will be recapitulated and a few new topics will be mentioned. New variants of the Reactive-tree should be considered that are able to deal efficiently with a non-hierarchical distribution of the number of objects over the importance levels, while sticking to the guideline that important objects are to be stored in the higher levels of the tree. This might be realised by allowing one tree level to contain multiple importance levels. Further, there are conceptual problems associated with scaleless geographic data sets, which have to be solved in co-operation with the cartographers.

The second topic requiring further research is the object management system of Persistent Procol. More practical experience has to be gained in developing GISs with Persistent Procol. This applies to both functional and efficiency-related issues. Furthermore, there are a few basic issues that still have to be solved, for example concurrent users of the same data set.

Other research topics are, of course, the ones that were outside the scope of this book (see p. *ix*), e.g. the development of a geographical query language and user-interface based on an object-oriented approach. The last future research topic mentioned deals with a geometric search tree that tries to improve response times by restructuring itself after a query, in a manner similar to a splay tree (Tarjan 1987) in the one-dimensional case. Basically, splay trees make entries that are often retrieved, easier to access by storing them in higher levels of the tree. If possible at all, the multi-dimensional self-adjusting tree will be more difficult.

Final Remark

Most proposed methods and techniques are also useful in other types of spatial information systems, such as computer graphics, computer-aided design and manufacturing (CAD/CAM), computer vision, and robot systems.

Appendix A: R-tree in Procol

This appendix contains the Procol code of the objects R-TREE and R-NODE, which together implement a persistent multi-dimensional index structure; see Section 10.5. In the case of a GIS with attributes such as points, polylines, and polygons in the plane, the dimension of the R-tree is two. A CAD system with solid objects, for example a polyhedron, will need a three-dimensional R-tree. Higher-dimensional R-trees are also possible, because in some applications it is beneficial to interpret a combination of k scalar attributes as one k-dimensional point attribute on which range queries may be formulated.

Not all the code is given here. Omitted code is indicated by three dots (...). In particular, some tricky parts of the Insert and Delete algorithms are omitted, but can be found in Guttman (1984). The code is non-trivial because the tree has to be kept in balance under the Insert and Delete operations. The purpose of this appendix is to show how the R-tree is implemented in an object-oriented manner. Only one query type is shown: the box select (in 2D, that is, a rectangle select), which returns the id of every object in the R-tree that overlaps the search box sbox. The result is sent back to the original object sid by invoking the action InRegion with the id of the found (graphic data) object. This happens once for each object that is found.

Note that we implement a parallel or distributed version of the tree by using the object R-NODE. During the Search operation several nodes can work in parallel on different processors. There is quite a bit of work in each node, because a typical value for M (maximum number of entries) is 100. This solution would probably not be very efficient for a binary tree, because the overhead introduced by sending messages to other processors may require more time than that gained by the parallel execution. Besides the Search operation, Delete and Insert operations might also benefit from parallel execution in the case of node overflow and underflow respectively. This is not shown in the code below.

```
#define DIM 2                    /* or any other value > 1 */
#define LENGTH 256

typedef struct{
    ANY      id;                 /* R-NODE or graphic object */
    float    box[DIM][2];
} EntryType;
```

```
-----------------------------------------------------------
OBJ         R-TREE (int m, int M, char dataset[LENGTH]);
Description In R-tree literature m and M are the minimum and
            maximum number of entries per node. The tree is
            suited for DIM-dimensional space. Assumed is that
            just after the creation of the R-TREE,
            it made persistent by its creator in 'dataset'.
Protocol    /* Assumption: The Add and Delete actions are */
            /* invoked by the (graphic data) objects       */
            /* themselves and they send their correct      */
            /* minimal bounding box in box.                */
            ANY (box)  -> Add      +
            ANY (box)  -> Delete   +
            ANY (sbox) -> Select
Declare     R-NODE      root, node;
            float       box[DIM][2], sbox[DIM][2];
Init        root = NULL;
Actions     Add = {
                if (root==NULL) {
                    new root(m, M, true);
                    persistent root in dataset;
                    root.EntryAdd(sender, box);
                } else {
                    ...
                    new node(m, M, true);
                    persistent node in dataset;
                    ...
                }
            }
            Delete = {
                ...
                /* If a (persistent) node is deleted,   */
                /* then this also implies removal from  */
                /* the data set                         */
                ...
            }
            Select = { root.Search(sender, box); }
EndOBJ      R-TREE.
-----------------------------------------------------------


-----------------------------------------------------------
OBJ         R-NODE (int m, int M, boolean leaf);
Description m and M have same meaning as in R-TREE.
            If leaf has value true, then this node is a leaf,
            else this is an internal node.
Protocol    (NrOfEntries < M): R-TREE (id, box)->EntryAdd   +
```

```
           (NrOfEntries > m): R-TREE (id, box)->EntryDelete+
           R-TREE (sid, sbox)                    ->Search    +
           R-NODE (sid, sbox)                     ->Search    +
           R-TREE ()                              ->Full
Declare    float         box[DIM][2], sbox[DIM][2];
           ANY           id, sid;
           EntryType     entry[M];
           int           NrOfEntries, i;
           R-NODE        next;
Init       NrOfEntries = 0;
Actions    EntryAdd = {
               /* Assumption: R-NODE is not full */
               entry[NrOfEntries].box  = box;
               entry[NrOfEntries++].id = id;
           }
           EntryDelete = { ... }
           Full = { sender.(NrOfEntries==M); }
           Search = {
               for (i = 0; i < NrOfEntries; i++)
                   if (overlap(sbox, entry[i].box)
                       if (leaf)
                           /* Return found object */
                           sid.InRegion(entry[i].id);
                       else {
                           /* Propagate search to lower       */
/* level. This part of the code   */
/* causes the parallelism.         */
                           next = entry[i].id;
                           next.Search(sid, sbox);
                       }
           }
EndOBJ     R-NODE.
----------------------------------------------------------------
```

Appendix B: Abbreviations

ADT	Abstract Data Type
AKR	Automatisering Kadastrale Registratie (Dutch)
AM/FM	Automated Mapping/Facility Management
ASCII	American Standard Code for Information Interchange
B-tree	Boeing/Bayer/balanced/broad/bushy-tree (Comer 1979)
BANG file	Balanced And Nested Grid file
BLG-tree	Binary Line Generalization-tree
BSP-tree	Binary Space Partitioning-tree
CAD	Computer Aided Design
CAM	Computer Aided Manufacturing
CARIN	CAR Information and Navigation system
C^3I	Command, Control, and Communication Information
CD-I	Compact Disk-Interactive
CEDD	Committee for the Exchange of Digital Data of the IHO
CGI	Computer Graphics Interface
CPU	Central Processing Unit
CSG	Constructive Solid Geometry
DBMS	DataBase Management System
DCDSTF	Digital Cartographic Data Standards Task Force
DDL	Data Definition Language
DFAD	Digital Feature Analysis Data
DIME	Dual Independent Map Encoding
DLG	Digital Line Graphs
DLMS	Digital LandMass System
DMA	Defense Mapping Agency (USA)
DML	Data Manipulation Language
DTED	Digital Terrain Elevation Data
EGR	Economic Geographic Region
EXCELL	Extendable CELL
FDS	Formal Data Structure
FEL	Fysisch en Elektronisch Laboratorium (Dutch)
GIS	Geographic Information System
GKS	Graphical Kernel System
GNU	GNU is Not Unix
hB-tree	holey brick B-tree

IDECAP	Interactive Pictorial Information System for Demographic and Environmental Planning Applications
IEEE	Institute of Electrical and Electronics Engineers
IGES	Initial Graphics Exchange Specification
IGU	International Geographical Union
IHO	International Hydrographic Organization
ISO	International Organization for Standardization
KD-tree	K-Dimensional tree (usually $K = 2$ in this book, and spelled as k-d tree by Bentley 1975)
KDB-tree	K-Dimensional B-tree (spelled as K-D-B tree by Robinson 1981)
KD2-tree	K-Dimensional tree with 2 split lines
KD2B-tree	K-Dimensional B-tree with 2 split lines
LIS	Land Information System
LKI	Landmeetkundig en Kartografisch Informatiesysteem (Dutch)
LSD tree	Local Split Decision tree
MACDIF	Map And Chart Data Interchange Format
MBE	Minimal Bounding Ellipse
MBR	Minimal Bounding Rectangle
MBS	Minimal Bounding Sphere
MSE	Mean Square Error
NATO	North Atlantic Treaty Organization
NCGIA	National Center for Geographic Information and Analysis
OMS	Object Management System
OO	Object-Oriented
OOPL	Object-Oriented Programming Language
OS	Operating System
PCTE+	Portable Common Tool Environment
PHIGS	Programmers' Hierarchical Interactive Graphics System
R-tree	Rectangle/range-tree (Guttman 1984)
RDBMS	Relational DataBase Management System
RNLAF	Royal NetherLands Air Force
SQL	Structured Query Language
SUF	Standaard Uitwisselings Formaat (Dutch)
TIGER	Topologically Integrated Geographic Encoding and Referencing
TNO	Nederlandse organisatie voor Toegepast-Natuurwetenschappelijk Onderzoek (Dutch)
USGS	United States Geological Survey
VLSI	Very Large Scale Integration
WDB I	World Data Bank I
WDB II	World Data Bank II

References

Abel, D. J. (1989). 'SIRO-DBMS: A Database Tool-kit for Geographical Information Systems', *International Journal of Geographical Information Systems*, *3(2)*, 103–16.

—— and Mark, D. M. (1990). 'A Comparative Analysis of Some Two-Dimensional Orderings', *International Journal of Geographical Information Systems*, *4(1)*, 21–31.

—— and Smith, J. L. (1983). 'A Data Structure and Algorithm Based on a Linear Key for a Rectangle Retrieval Problem', *Computer Vision, Graphics and Image Processing*, *24*, 1–13.

Adobe Systems Incorporated (1985). *PostScript Language Reference Manual*. Addison-Wesley, Reading, Mass.

—— (1987). *PostScript Language Tutorial and Cookbook*. Addison-Wesley, Reading, Mass.

Anderson, N. (1989). 'Review of the Development of Digital Geographical Interchange Standards', in *Digital Data Exchange Seminar held at the International Hydrographic Bureau, Monaco, November 1988*, pp. 115–21 Monaco. International Hydrographic Bureau.

Andriole, S. J. (1986). 'Graphic Equivalence, Graphic Explanations and Embedded Process Modeling for Enhanced User-System Interaction', *IEEE Transactions on Systems, Man and Cybernetics*, *SMC-16(6)*, 919–29.

Annoni, A., Ventura, A. D., Mozzi, E., and Schettini, R. (1990). 'Towards the Integration of Remote Sensing Images within a Cartographic System', *Computer-Aided Design*, *22(3)*, 160–6.

ANSI (1985). 'Computer Graphics – Graphical Kernel System (GKS) – Functional Description', Tech. rep. ANSI X3.124-1985, American National Standards Institute.

Atkinson, M. P., and Buneman, O. P. (1987). 'Types and Persistence in Database Programming Languages', *ACM Computing Surveys*, *19(2)*, 105–90.

—— Bailey, P. J., Chisholm, K. J., Cockshott, P. W., and Morrison, R. (1983). 'An Approach to Persistent Programming', *Computer Journal*, *26(4)*, 360–5.

Bakker, G., de Munck, J. C., and van Hees, G. L. S. (1986). 'Radio Positioning on Sea', Faculty of Geodesy, Delft University of Technology.

Ballard, D. H. (1981). 'Strip Trees: A Hierarchical Representation for Curves', *Communications of the ACM*, *24(5)*, 310–21.

Bancilhon, F., Barbedette, G., Benzaken, V., Delobel, C., Gamerman, S., Lécluse, C., Pfeffer, P., Richard, P., and Velez, F. (1988). 'The Design and Implementation of O_2, an Object-Oriented Database System', in *Advances in Object-Oriented Database Systems: 2nd International Workshop on Object-Oriented Database Systems,* Bad Münster am Stein-Ebernburg, FRG (Berlin, Springer-Verlag), pp. 1–22.

Barker, G. R. (1988). 'Remote Sensing: The Unheralded Component of Geographic Information Systems', *Photogrammetric Engineering and Remote Sensing, 54(2),* 195–9.

Bayer, R., and McCreight, E. (1973). 'Organization and Maintenance of Large Ordered Indexes', *Acta Informatica, 1,* 173–89.

Becker, B., Six, H.-W., and Widmayer, P. (1991). 'Spatial Priority Search: An Access Technique for Scaleless Maps', *ACM SIGMOD, 20(2),* 128–37.

Beckley, D. A., Evens, M. W., and Raman, V. K. (1985). 'Multikey Retrieval from KD-trees and Quad Trees', *ACM SIGMOD, 14(4),* 291–301.

Beckmann, N., Kriegel, H.-P., Schneider, R., and Seeger, B. (1990). 'The R^*-tree: An Efficient and Robust Access Method for Points and Rectangles', in *ACM/SIGMOD, Atlantic City* (New York, ACM), pp. 322–31.

Bennis, K., David, B., Quilio, I., and Viémont, Y. (1990). 'GéoTropics Database Support Alternatives for Geographic Applications', in *4th International Symposium on Spatial Data Handling, Zürich* (Columbus, OH, International Geographical Union), pp. 599–610.

—— —— Morize-Quilio, I., Thévenin, J. M., and Viémont, Y. (1991). 'GéoGraph: A Topological Storage Model for Extensible GIS', in *Auto-Carto 10,* pp. 349–67.

Bentley, J. L. (1975). 'Multidimensional Binary Search Trees Used for Associative Searching', *Communications of the ACM, 18(9),* 509–17.

—— and Friedman, J. H. (1979). 'Data Structures for Range Searching', *Computing Surveys, 11(4),* 397–409.

—— and Ottmann, T. A. (1979). 'Algorithms for Reporting and Counting Geometric Intersections', *IEEE Transactions on Computers, C-28(9),* 643–7.

Berman, R. R., and Stonebraker, M. (1977). 'GEO-QUEL: A System for the Manipulation and Display of Geographic Data', *ACM Computer Graphics, 11(2),* 186–91.

Blake, E. H., and Cook, S. (1987). 'On Including Part Hierarchies in Object-Oriented Languages, with an Implementation in Smalltalk', in *ECOOP '87,* pp. 41–50.

Blankenagel, G., and Güting, R. H. (1988). 'Internal and External Algorithms for the Points-in-Regions Problem: The Inside Join of Geo-Relational Algebra', in *CG'88: International Workshop on Computational Geometry* (Berlin, Springer-Verlag), pp. 85–9.

Boudriault, G. (1987). 'Topology in the TIGER File', in *Auto-Carto 8*, pp. 258–69.

Brassel, K. E., and Weibel, R. (1988). 'A Review and Conceptual Framework of Automated Map Generalization', *International Journal of Geographical Information Systems*, *2(3)*, 229–44.

Buisson, L. (1989). 'Reasoning on Space with Object-Centered Knowledge Representation', in *Symposium on the Design and Implementation of Large Spatial Databases, Santa Barbara, California* (Berlin, Springer-Verlag), pp. 325–44.

Burrough, P. A. (1986). *Principles of Geographical Information Systems for Land Resources Assessment.* Monographs on Soil and Resources Survey no. 12. Oxford University Press.

—— (1987). 'Multiple Sources of Spatial Variation and How to Deal with them', in *Auto-Carto 8*, pp. 145–54.

Carter, J. R. (1989). 'Relative Errors Identified in USGS Gridded DEMs', in *Auto-Carto 9*, pp. 255–65.

Chance, A., Newell, R. G., and Theriault, D. G. (1990). 'An Object-Oriented GIS: Issues and Solutions', in *EGIS'90: First European Conference on Geographical Information Systems, Amsterdam, the Netherlands* (Utrecht, EGIS Foundation), pp. 179–88.

Chang, N. S., and Fu, K. S. (1980). 'A Relational Database System for Images', in *Pictorial Information Systems*, Lecture Notes in Computer Science no. 80, pp. 288–321.

Chang, S.-K., and Kunii, T. L. (1981). 'Pictorial Data-Base Systems', *Computer (USA)*, *14(11)*, 13–21.

Chazelle, B., and Edelsbrunner, H. (1992). 'An Optimal Algorithm for Intersecting Line Segments in the Plane', *Journal of the Association for Computing Machinery*, *39(1)*, 1–54.

Chen, P. P.-S. (1976). 'The Entity-Relationship Model: Toward a Unified View of Data', *ACM Transactions on Database Systems*, *1(1)*, 9–36.

Chrisman, N. R. (1987). 'Efficient Digitizing through the Combination of Appropriate Hardware and Software for Error Detection and Editing', *International Journal of Geographical Information Systems*, *1(3)*, 265–77.

—— (1990a). 'Deficiencies of Sheets and Tiles: Building Sheetless Databases', *International Journal of Geographical Information Systems*, *4(2)*, 157–67.

—— (1990b). 'Spatial Indexing Schemes for GIS: A Critique', University of Washington.

Claassen, E. (1989). 'Spatial Data Structures in Geographic Information Systems', Master's thesis, Department of Computer Science, University of Leiden.

Clementini, E., and Felice, P. D. (1990). 'An Extensible Class Library for Geographic Applications', in *TOOLS'90* (Paris, Editions Angkor), pp. 613–23.

Clementini, E., D'Atri, A., and Felice, P. D. (1990). 'Browsing in Geographic Databases: An Object-Oriented Approach', in *IEEE Workshop on Visual Languages, Skokie, Illinois* (Los Alamitos, CA, Computer Society Press), pp. 125–31.

Cockshott, W. P. (1987). 'Persistent Programming and Secure Data Storage', *Information and Software Technology*, *29(5)*, 249–56.

Codd, E. F. (1970). 'A Relational Model of Data for Large Shared Data Banks', *Communications of the ACM*, *13(6)*, 377–87.

Comer, D. (1979). 'The Ubiquitous B-Tree', *ACM Computing Surveys*, *11(2)*, 121–37.

Cowen, D. J. (1988). 'GIS versus CAD versus DBMS: What are the Differences?', *Photogrammetric Engineering and Remote Sensing*, *54(11)*, 1551–5.

Cox, B. J. (1986). *Object-Oriented Programming: An Evolutionary Approach*. Addison-Wesley, Reading, Mass.

Date, C. J. (1981). *An Introduction to Database Systems*. Addison-Wesley, Reading, Mass.

DCDSTF (1988). 'Digital Cartographic Data Standards Task Force: The Proposed Standard for Digital Cartographic Data', *American Cartographer*, *15(1)*. Special Issue.

DGIWG (1991). 'DIGEST – Digital Geographic Information – Exchange Standards – Edition 1.0', Defence Mapping Agency, USA, Digital Geographic Information Working Group.

Dietrich, Jr, W. C., Nackman, L. R., Sundaresan, C. J., and Gracer, F. (1989). 'TGMS: An Object-Oriented System for Programming Geometry', *Software – Practice and Experience*, *19(10)*, 979–1013.

DMA (1977). 'Product Specifications for Digital Landmass System (DLMS) Data Base', Defense Mapping Agency, Aerospace Center, St Louis, Mo.

——— (1986*a*). 'Product Specifications for Digital Feature Analysis Data (DFAD): Level 1 and Level 2', Defense Mapping Agency, Aerospace Center, St Louis, Mo.

——— (1986*b*). 'Product Specifications for Digital Terrain Elevation Data (DTED)', Defense Mapping Agency, Aerospace Center, St Louis, Mo.

Doerschler, J. S. (1987). *A Rule-Based System for Dense-Map Name Placement*. Ph.D. thesis, Rensselaer Polytechnic Institute, Troy, NY.

Dougenik, J. (1979). 'WHIRLPOOL: A Geometric Processor for Polygon Coverage Data', in *Auto-Carto 4*, pp. 304–11.

Douglas, D. H., and Peucker, T. K. (1973). 'Algorithms for the Reduction of Points Required to Represent a Digitized Line or its Caricature', *Canadian Cartographer*, *10*, 112–22.

Doytsher, Y., and Shmutter, B. (1986). 'Intersecting Layers of Information: A Computerized Solution', in *Auto-Carto London*, pp. 136–45.

Duda, R. O., and Hart, P. E. (1973). *Pattern Classification and Scene Analysis*. John Wiley, New York.

Edelsbrunner, H., and Welzl, E. (1986). 'Halfplanar Range Search in Linear Space and $O(n^{0.695})$ Query Time', *Information Processing Letters, 23*, 289–93.

—— (1987). *Algorithms in Combinatorial Geometry*. EATCS Monographs on Theoretical Computer Science. Springer-Verlag, Berlin.

Egenhofer, M. J., and Frank, A. U. (1988). 'Designing Object-Oriented Query Languages for GIS: Human Interface Aspects', in *Proceedings of the 3rd International Symposium on Spatial Data Handling* (Columbus, OH, International Geographical Union), pp. 79–96.

—— —— (1989a). 'PANDA: An Extensible DBMS Supporting Object-Oriented Software Techniques', in *Database Systems in Office, Engineering, and Scientific Environment* New York.

—— —— (1989b). 'Object-Oriented Modeling in GIS: Inheritance and Propagation', in *Auto-Carto 9*, pp. 588–98.

Elzinga, J., and Hearn, D. W. (1972). 'Geometrical Solutions for Some Minimax Location Problems', *Transportation Science, 6*, 379–94.

ESRI (1989). 'ARC/INFO: The Georelational Model Revisited', *ARC News, 11(1)*, 9.

Evenden, G. I. (1990). 'Cartographic Projection Procedures for the UNIX Environment: A User's Manual', Tech. rep. Open-File Report 90–284, US Department of the Interior, Geological Survey, Woods Hole, Mass.

Falcidieno, B., and Spagnuolo, M. (1991). 'A New Method for the Characterization of Topographic Surfaces', *International Journal of Geographical Information Systems, 5(4)*, 397–412.

Faloutsos, C. (1988). 'Gray Codes for Partial Match and Range Queries', *IEEE Transactions on Software Engineering, SE-14(10)*, 1381–93.

—— Sellis, T., and Roussopoulos, N. (1987). 'Analysis of Object Oriented Spatial Access Methods', *ACM SIGMOD, 16(3)*, 426–39.

Feeley, J., and O'Brien, D. (1989). 'MACDIF Map and Chart Data Interchange Format', in *Digital Data Exchange Seminar* held at the International Hydrographic Bureau, Monaco, November 1988 (Monaco, International Hydrographic Bureau), pp. 106–14.

Fisher, P. F., and Lindenberg, R. E. (1989). 'On Distinctions among Cartography, Remote Sensing, and Geographic Information Systems', *Photogrammetric Engineering and Remote Sensing, 55(10)*, 1431–4.

Foley, J. D., and van Dam, A. (1982). *Fundamentals of Interactive Computer Graphics*. Addison-Wesley, Reading, Mass.

Frank, A. (1981). 'Application of DBMS to Land Information Systems', in *Proceedings of the Seventh International Conference on Very Large Data Bases* (New York, ACM), pp. 448–53.

—— (1982). 'MAPQUERY: Data Base Query Language for Retrieval of Geometric Data and their Graphical Representation', *ACM Computer Graphics, 16(3)*, 199–207.

—— (1983). 'Storage Methods for Space Related Data: The Field-tree', Tech. rep. Bericht no. 71, Eidgenössische Technische Hochschule Zürich.

Frank, A. (1987). 'Overlay Processing in Spatial Information Systems', in *Auto-Carto 8*, pp. 12–31.

—— (1988a). 'Multiple Inheritance and Genericity for the Integration of a Database Management System in an Object-Oriented Approach', in *Advances in Object-Oriented Database Systems: 2nd International Workshop on Object-Oriented Database Systems,* Bad Münster am Stein-Ebernburg, FRG (Berlin, Springer-Verlag), pp. 268–73.

—— (1988b). 'Requirements for a Database Management System for a GIS', *Photogrammetric Engineering and Remote Sensing, 54(11),* 1557–64.

—— and Barrera, R. (1989). 'The Field-tree: A Data Structure for Geographic Information System', in *Symposium on the Design and Implementation of Large Spatial Databases, Santa Barbara, California* (Berlin, Springer-Verlag), pp. 29–44.

Franklin, W. R. (1990). 'Calculating Map Overlay Polygons' Areas without Explicitly Calculating the Polygons: Implementation', in *4th International Symposium on Spatial Data Handling, Zürich* (Columbus, OH, International Geographical Union), pp. 151–60.

—— and Lewis, H. L. (1978). '3D Graphic display of Discrete Spatial Data by PRISM Maps', *ACM Computer Graphics, 12(3),* 70–5.

—— and Wu, P. Y. F. (1987). 'A Polygon Overlay System in PROLOG', in *Auto-Carto 8*, pp. 97–106.

Freeman, H., and Ahn, J. (1984). 'AUTONAP: An Expert System for Automatic Map Name Placement', in *Proceedings 1st International Symposium on Spatial Data Handling, Zürich* (Columbus, OH, International Geographical Union), pp. 544–71.

—— —— (1987). 'On the Problem of Placing Names in a Geographic Map', *International Journal of Pattern Recognition and Artificial Intelligence, 1(1),* 121–40.

Freeston, M. (1987). 'BANG File: A New Kind of Grid File', *ACM SIGMOD, 16(4),* 260–9.

—— (1989). 'A Well-Behaved File Structure for the Storage of Spatial Objects', in *Symposium on the Design and Implementation of Large Spatial Databases, Santa Barbara, California* (Berlin, Springer-Verlag), pp. 286–300.

Fuchs, H., Kedem, Z. M., and Naylor, B. F. (1980). 'On Visible Surface Generation by A Priori Tree Structures', *ACM Computer Graphics, 14(3),* 124–33.

—— Abram, G. D., and Grant, E. D. (1983). 'Near Real-Time Shaded Display of Rigid Objects', *ACM Computer Graphics, 17(3),* 65–72.

Gahegan, M. N., and Roberts, S. A. (1988). 'An Intelligent, Object-Oriented Geographical Information System', *International Journal of Geographical Information Systems, 2(2),* 101–10.

Goh, P.-C. (1989). 'A Graphic Query Language for Cartographic and Land Information Systems', *International Journal of Geographical Information Systems, 3(3),* 245–55.

Goodchild, M. F. (1987*a*). 'A Model of Error for Choropleth Maps, with Applications to Geographic Information Systems', in *Auto-Carto 8*, pp. 165–74.

―――― (1987*b*). 'A Spatial Analytical Perspective on Geographical Information Systems', *International Journal of Geographical Information Systems, 1(4)*, 327–34.

―――― (1989). 'Tiling Large Geographical Databases', in *Symposium on the Design and Implementation of Large Spatial Databases, Santa Barbara, California* (Berlin, Springer-Verlag), pp. 137–46.

―――― and Grandfield, A. W. (1983). 'Optimizing Raster Storage: An Examination of Four Alternatives', in *Auto-Carto 6*, pp. 400–7.

―――― and Min-hua, W. (1988). 'Modeling Error in Raster-Based Spatial Data', in *Proceedings of the 3rd International Symposium on Spatial Data Handling* (Columbus, OH, International Geographical Union), pp. 97–106.

Goodenough, D. G. (1988). 'Thematic Mapper and SPOT Integration with a Geographic Information System', *Photogrammetric Engineering and Remote Sensing, 54(2)*, 167–76.

Gorny, A. J., and Carter, R. (1987). 'World Data Bank II: General Users Guide', US Central Intelligence Agency.

Greene, D. (1989). 'An Implementation and Performance Analysis of Spatial Data Access Methods', in *Proceedings IEEE Fifth International Conference on Data Engineering, Los Angeles, California* (Los Alamitos, CA, IEEE Computer Society Press), pp. 606–15.

Greenlee, D. D. (1987). 'Raster and Vector Processing for Scanned Linework', *Photogrammetric Engineering and Remote Sensing, 53(10)*, 1383–7.

Günther, O. (1988). *Efficient Structures for Geometric Data Management*. in Lecture Notes in Computer Science no. 337. Springer-Verlag, Berlin.

―――― and Bilmes, J. (1988). 'The Implementation of the Cell-tree: Design Alternatives and Performance Evaluation', Tech. rep. TRCS88-23, University of California, Santa Barbara.

―――― ―――― (1991). 'Tree-Based Access Methods for Spatial Databases: Implementation and Performance Evaluation', *IEEE Transactions on Knowledge and Data Engineering, 3(3)*, 342–56.

Guptill, S. C. (1988). 'A Process for Evaluation Geographic Information Systems', Tech. rep. Open-File Report 88–105, US Geological Survey, Federal Interagency Coordinating Committee on Digital Cartography.

―――― (1989*a*). 'Evaluating Geographic Information Systems Technology', *Photogrammetric Engineering and Remote Sensing, 55(11)*, 1583–7.

―――― (1989*b*). 'Speculations on Seamless, Scaleless Cartographic Data Bases', in *Auto-Carto 9*, pp. 436–43.

Güting, R. H. (1988*a*). 'Geo-Relational Algebra: A Model and Query Language for Geometric Database Systems', in *CG'88: International Workshop on Computational Geometry* (Berlin, Springer-Verlag), pp. 90–6.

Güting, R. H. (1988*b*). 'Geo-Relational Algebra: A Model and Query Language for Geometric Database Systems', in *Advances in Database Technology: EDBT'88* (Berlin, Springer-Verlag), pp. 506–27.

Guttman, A. (1984). 'R-Trees: A Dynamic Index Structure for Spatial Searching', *ACM SIGMOD, 13*, 47–57.

Henrich, A., Six, H.-W., and Widmayer, P. (1989). 'The LSD Tree: Spatial Access to Multidimensional Point and Non-Point Objects', in *Proceedings of the Fifteenth International Conference on Very Large Data Bases, Amsterdam* (New York, ACM), pp. 45–53.

Herring, J. R. (1987). 'TIGRIS: Topologically Integrated Geographic Information System', in *Auto-Carto 8*, pp. 282–91.

Heuvelink, G. B. M., Burrough, P. A., and Stein, A. (1989). 'Propagation of Errors in Spatial Modelling with GIS', *International Journal of Geographical Information Systems, 3(4)*, 303–22.

Hutflesz, A., Widmayer, P., and Six, H.-W. (1988). 'Twin Grid Files: A Performance Evaluation', in *CG'88: International Workshop on Computational Geometry* (Berlin, Springer-Verlag), pp. 15–24.

—— Six, H.-W., and Widmayer, P. (1990). 'The R-File: An Efficient Access Structure for Proximity Queries', in *Proceedings IEEE Sixth International Conference on Data Engineering, Los Angeles, California* (Los Alamitos, CA, IEEE Computer Society Press), pp. 372–9..

IEPG TA-13 (1988). 'PCTE+ C Functional Specification Issue 2', Independent European Programme Group, Technical Area 13.

—— (1989). 'Introducing PCTE+', Independent European Programme Group, Technical Area 13.

Ingram, K. J., and Phillips, W. W. (1987). 'Geographic Information Processing using a SQL-based Query Language', in *Auto-Carto 8*, pp. 326–35.

Intergraph (1990). 'MGE: The Modular GIS Environment', brochure.

ISO (1985). 'Computer Graphics – Graphical Kernel System for Three Dimensions (GKS-3D) – Functional Description', Tech. rep. ISO DP 8805, International Organization for Standardization.

—— (1986*a*). 'Information Processing Systems – Computer Graphics – Interfacing Techniques for Dialogues with Graphical Devices – Functional Specification CGI', Tech. rep. ISO DP 9636, International Organization for Standardization.

—— (1986*b*). 'Information Processing Systems – Computer Graphics – Metafile for the Storage and Transfer of Picture Description Information', Tech. rep. ISO DIS 8632, International Organization for Standardization.

—— (1989*a*). 'Information Processing Systems – Computer Graphics – Programmers Hierarchical Interactive Graphics System (PHIGS) – Part 1: Functional Description', Tech. rep. ISO IEC 9592-1, International Organization for Standardization.

ISO (1989*b*). 'Information Processing Systems – Computer Graphics – Programmers Hierarchical Interactive Graphics System (PHIGS) – Part 4: Plus Lumière und Surfaces (PHIGS PLUS)', Tech. rep. ISO IEC 9592-4, International Organization for Standardization.

Jagadish, H. V. (1990). 'Linear Clustering of Objects with Multiple Attributes', in *ACM/SIGMOD, Atlantic City* (New York, ACM), pp. 332–42.

Jense, H. (1991). *Interactive Inspection of Volume Data*. Ph.D. thesis, Department of Computer Science, Leiden University.

Jones, C. B. (1989). 'Data Structures for Three-Dimensional Spatial Information Systems in Geology', *International Journal of Geographical Information Systems*, *3(1)*, 15–31.

―――― (1990). 'Conflict Resolution in Cartographic Name Placement', *Computer-Aided Design*, *22(3)*, 173–83.

―――― and Abraham, I. M. (1986). 'Design Considerations for a Scale-Independent Cartographic Database', in *Proceedings 2nd International Symposium on Spatial Data Handling, Seattle* (Columbus, OH, International Geographical Union), pp. 348–98.

―――― ―――― (1987). 'Line Generalization in a Global Cartographic Database', *Cartographica*, *24(3)*, 32–45.

Joseph, T., and Cardenas, A. F. (1988). 'PICQUERY: A High Level Query Language for Pictorial Database Management', *IEEE Transactions on Software Engineering*, *14(5)*, 630–8.

Kemper, A., and Wallrath, M. (1987). 'An Analysis of Geometric Modeling in Database Systems', *ACM Computing Surveys*, *19(1)*, 47–91.

Kernighan, B. W., and Ritchie, D. M. (1978). *The C Programming Language*. Prentice-Hall, Englewood Cliffs, NJ.

Khoshafian, S. N., and Copeland, G. P. (1986). 'Object Identity', in *OOPSLA'86* (New York, ACM), pp. 406–16.

Kinnear, C. (1987). 'The TIGER Structure', in *Auto-Carto 8*, pp. 249–57.

Kleiner, A., and Brassel, K. E. (1986). 'Hierarchical Grid Structures for Static Geographic Data Bases', in *Auto-Carto London*, pp. 485–96.

Knuth, D. E. (1973). *Sorting and Searching*, vol. 3 of *The Art of Computer Programming*. Addison-Wesley, Reading, Mass.

Konečný, M. (ed.). (1991). *IGU Brno GIS'91 Conference*, Columbus, OH. International Geographical Union, Commission on GIS.

Kraak, M. J. (1986). 'Computer Assisted Cartographic Three-Dimensional Imaging Techniques', in *Auto-Carto London*, pp. 53–8.

―――― (1988). *Computer-Assisted Cartographical Three-Dimensional Imaging Techniques*. Ph.D. thesis, Delft University of Technology.

Kriegel, H.-P., Schiwiets, M., Schneider, R., and Seeger, B. (1989). 'Performance Comparison of Point and Spatial Access Methods', in *Symposium on the Design and Implementation of Large Spatial Databases, Santa Barbara, California* (Berlin, Springer-Verlag), pp. 89–114.

Kriegel, H.-P., Brinkhoff, T., and Schneider, R. (1991). 'The Combination of Spatial Access Methods and Computational Geometry in Geographic Database Systems', in *Advances in Spatial Databases, 2nd Symposium, SSD'91, Zürich* (Berlin, Springer-Verlag), pp. 3–21.

Laffra, C., and van Oosterom, P. (1991). 'Persistent Graphical Objects', in *Advances in Object-Oriented Graphics 1: First Eurographics Workshop on Object Oriented Graphics, June 1990, Königswinter, FRG* Berlin. Springer-Verlag.

Lahaije, P. D. M. E. (1986). 'Algorithms and Data Structures for Efficient Retrieval of CARIN Roadmap Data from a Compact Disc ROM', Master's thesis, Eindhoven University of Technology.

Lee, J. (1991*a*). 'Analysis of Visibility Sites on Topographic Surfaces', *International Journal of Geographical Information Systems, 5 (4)*, 413–29.

—— (1991*b*). 'Comparison of Existing Methods for Building Triangular Network Models of Terrain From Grid Digital Elevation Models', *International Journal of Geographical Information Systems, 5 (3)*, 267–85.

Logan, T. L., and Bryant, N. A. (1987). 'Spatial Data Software Integration: Merging CAD/ CAM/ Mapping with GIS and Image Processing', *Photogrammetric Engineering and Remote Sensing, 53 (10)*, 1391–5.

Lohrenz, M. C. (1988). 'Design and Implementation of the Digital Vector Shoreline Data Format', Tech. rep. 194, Naval Ocean Research and Development Activity.

Lomet, D. B., and Salzberg, B. (1989). 'A Robust Multi-Attribute Search Structure', in *Proceedings IEEE Fifth International Conference on Data Engineering, Los Angeles, California* (Los Alamitos, CA, IEEE Computer Society Press), pp. 296–304.

Lupien, A. E., Moreland, W. H., and Dangermond, J. (1987). 'Network Analysis in Geographic Information Systems', *Photogrammetric Engineering and Remote Sensing, 53 (10)*, 1417–21.

Mairson, H. G., and Stolfi, J. (1988). 'Reporting and Counting Intersections between Two Sets of Line Segments', in *Theoretical Foundations of Computer Graphics and CAD*, vol. 40 of *NATO ASI Series F* (Berlin, Springer-Verlag), pp. 307–25.

Mäntylä, M. (1988). *An Introduction to Solid Modeling.* Computer Science Press, Rockville, Md.

Mark, D. M. (1989). 'Conceptual Basis for Geographic Line Generalization', in *Auto-Carto 9*, pp. 68–77.

Marx, R. W. (1986). 'The TIGER System: Automating the Geographic Structure of the United States Census', *Government Publications Review, 13*, 181–201.

Matsuyama, T., Hao, L. V., and Nagao, M. (1984). 'A File Organization for Geographic Information Systems Based on Spatial Proximity', *Computer Vision, Graphics and Image Processing, 26*, 303–18.

McMaster, R. B. (1987). 'Automated Line Generalization', *Cartographica*, *24(2)*, 74–111.

Megiddo, N. (1983). 'Linear-Time Algorithms for Linear Programming in R^3 and Related Problems', *SIAM Journal on Computing*, *12(4)*, 759–76.

Menon, S., and Smith, T. R. (1989). 'A Declarative Spatial Query Processor for Geographic Information Systems', *Photogrammetric Engineering and Remote Sensing*, *55(11)*, 1593–1600.

Mercera, P. A. (1991). 'A Geometric Extension to Postgres', Tech. rep. FEL-91-S304, FEL-TNO Divisie 2.

Meyer, B. (1988). *Object-oriented Software Construction*. Prentice-Hall, Hemel Hempstead.

Mohan, L., and Kashyap, R. L. (1988). 'An Object-Oriented Knowledge Representation for Spatial Information', *IEEE Transactions on Software Engineering*, *14(5)*, 675–81.

Molenaar, M. (1989). 'Single Valued Vector Maps: A Concept in Geographic Information Systems', *Geo-Informationssysteme*, *2(1)*, 18–26.

—— (1990). 'A Formal Data Structure for Three Dimensional Vector Maps', in *4th International Symposium on Spatial Data Handling, Zürich* (Columbus, OH, International Geographical Union), pp. 830–43.

Mood, A. M., Graybill, F. A., and Boes, D. C. (1974). *Introduction to the Theory of Statistics*. McGraw-Hill, Tokyo.

Morehouse, S. (1989). 'The Architecture of Arc/Info', in *Auto-Carto 9*, pp. 266–77.

Moreland, W. H., and Lupien, A. E. (1987). 'Realistic Flow Analysis Using a Simple Network Model', in *Auto-Carto 8*, pp. 122–8.

Morrison, J. (1989). 'US Digital Cartographic Data Standards: Spatial Data Transfer Specifications (SDTS)', in *Digital Data Exchange Seminar held at the International Hydrographic Bureau, Monaco, November 1988* (Monaco, International Hydrographic Bureau), pp. 89–95.

Morrison, R., Florianis, A. L., Dearle, A., and Atkinson, M. P. (1986). 'An Integrated Graphics Programming Environment', *Computer Graphics Forum*, *5*, 147–57.

Mortenson, M. E. (1985). *Geometric Modeling*. John Wiley, New York.

Müller, J.-C. (1987). 'Optimum Point Density and Compaction Rates for the Representation of Geographic Lines', in *Auto-Carto 8*, pp. 221–30.

—— (1990*a*). 'Generalization of Spatial Data Bases', ITC Enschede, Netherlands.

—— (1990*b*). 'Rule Based Generalization: Potentials and Impediments', in *4th International Symposium on Spatial Data Handling, Zürich* (Columbus, OH, International Geographical Union), pp. 317–34.

—— (1990*c*). 'Rule Based Selection for Small Scale Map Generalization', ITC Enschede, Netherlands.

Nagel, R. N., Braithwaite, W. W., and Kennicott, P. R. (1980). 'Initial Graphics Exchange Specification, IGES Version 1.0', National Bureau of Standards, Washington, DC.

Nagy, G., and Wagle, S. (1979). 'Geographic Data Processing', *Computer Surveys*, *11(2)*, 139–81.

NCGIA, T. (1989). 'Research Initiative 3: Multiple Representations', *NCGIA Update*, *1(2)*, 2.

Newman, W. W., and Sproull, R. F. (1981). *Principles of Interactive Computer Graphics* (2nd ed.),. McGraw-Hill, New York.

Nickerson, B. G. (1987). *Automatic Cartographic Generalization For Linear Features*. Ph.D. thesis, Rensselaer Polytechnic Institute, Troy, NY.

—— and Freeman, H. (1986). 'Development of a Rule-Based System for Automatic Map Generalization', in *Proceedings 2nd International Symposium on Spatial Data Handling, Seattle* (Columbus, OH, International Geographical Union), pp. 537–56.

Nievergelt, J., Hinterberger, H., and Sevcik, K. C. (1984). 'The Grid File: An Adaptable, Symmetric Multikey File Structure', *ACM Transactions on Database Systems*, *9(1)*, 38–71.

NSF (1987). 'The National Center for Geographic Information and Analysis: A Prospectus', National Science Foundation.

Ooi, B. C. (1990). *Efficient Query Processing in Geographic Information Systems*. Lecture Notes in Computer Science no. 471. Springer-Verlag, Berlin.

—— McDonell, K. J., and Sacks-Davis, R. (1987). 'Spatial KD-Tree: An Indexing Mechanism for Spatial Database', in *Proceedings IEEE COMPSAC'87: The Eleventh Annual International Computer Software & Applications Conference, Tokyo* (Los Alamitos, CA, IEEE Computer Society Press), pp. 433–8.

—— Sacks-Davis, R., and McDonell, K. J. (1989). 'Extending a DBMS for Geographic Applications', in *Proceedings IEEE Fifth International Conference on Data Engineering, Los Angeles, California* (Los Alamitos, CA, IEEE Computer Society Press), pp. 590–7.

Oracle Corporation (1990). *SQL Language Reference Manual, Version 6.0.* Oracle Corporation, Redwood Shores, Cal.

Orenstein, J. (1991). 'An Algorithm for Computing the Overlay of k-Dimensional Spaces', in *Advances in Spatial Databases, 2nd Symposium, SSD'91, Zürich* (Berlin, Springer-Verlag), pp. 381–400.

—— and Manola, F. A. (1988). 'PROBE Spatial Data Modeling and Query Processing in an Image Database Application', *IEEE Transactions on Software Engineering*, *14(5)*, 611–29.

Ormeling, F. J., and Kraak, M. J. (1987). *Kartografie: Ontwerp, productie en gebruik van kaarten* (in Dutch). Delft University Perss.

Overmars, M. H. (1983). *The Design of Dynamic Data Structures*. Lecture Notes in Computer Science no. 156. Springer-Verlag, Berlin.

Paterson, M. S., and Yao, F. F. (1989). 'Binary Partitions with Applications to Hidden-Surface Removal and Solid Modelling', in *Proceedings 5th ACM Symposium on Computational Geometry* (New York, ACM), pp. 23–32.

Paton, N. W., and Gray, P. M. D. (1988). 'Identification of Database Objects by Key', in *Advances in Object-Oriented Database Systems: 2nd International Workshop on Object-Oriented Database Systems*, Bad Münster am Stein-Ebernburg, FRG (Berlin, Springer-Verlag), pp. 280–5.

Pedersen, L.-O., and Spooner, R. (1988). 'Data Organization in System 9', WILD Heerbrugg.

Penney, D. J., and Stein, J. (1987). 'Class Modification in the GemStone Object-Oriented DBMS', in *OOPSLA '87* (New York, ACM), pp. 111–17.

Persson, J., and Jungert, E. (1991). 'Generation of Hierarchical Multi-resolution Maps Represented in Nonhierarchical Map Data Structure', in *Proceedings EGIS'91: Second European Conference on Geographical Information Systems* (Utrecht, EGIS Foundation), pp. 866–75.

Petrie, G., and Kennie, T. J. M. (eds.). (1990). *Terrain Modelling in Surveying and Civil Engineering*. Whittles Publishing, Latheronwheel, Caithness.

Peucker, T. K., and Chrisman, N. (1975). 'Cartographic Data Structures', *American Cartographer*, *2(1)*, 55–69.

Peuquet, D. J. (1981*a*). 'An Examination of Techniques for Reformatting Digital Cartographic Data, Part 1: The Raster-to-Vector Process', *Cartographica*, *18(1)*, 34–48.

—— (1981*b*). 'An Examination of Techniques for Reformatting Digital Cartographic Data, Part 2: The Vector-to-Raster Process', *Cartographica*, *18(3)*, 21–33.

—— (1983). 'A Hybrid Structure for the Storage and Manipulation of Very Large Spatial Data Sets', *Computer Vision, Graphics and Image Processing*, *24*, 14–27.

—— (1984). 'Data Structures for a Knowledge-Based Geographic Information System', in *Proceedings of the International Symposium on Spatial Data Handling, Zürich* (Columbus, OH, International Geographical Union), pp. 371–91.

Postgres Research Group (1991). 'The Postgres Reference Manual, Version 3.1', Tech. rep. Memorandum, Electronics Research Laboratory, College of Engineering.

Preparata, F. P., and Shamos, M. I. (1985). *Computational Geometry*. Springer-Verlag, New York.

Press, W. H., Flannery, B. P., Teukolsky, S. A., and Vetterling, W. T. (1988). *Numerical Recipes in C: The Art of Scientific Computing*. Cambridge University Press.

Projectgroep (1982). 'IDECAP Interactief Pictorieel Informatiesysteem voor Demografische en Planologische Toepasingen: Een verkennend en vergelijkend onderzoek', Tech. rep. Publicatiereeks 1982/2, Stichting Studiecentrum voor Vastgoedinformatie te Delft.

Puk, R. F., and McConnell, J. I. (1986). 'GKS-3D: A Three-Dimensional Extension to the Graphical Kernel System', *IEEE Computer Graphics & Applications*, *6(8)*, 42–9.

Pullar, D. (1991). 'Spatial Overlay with Inexact Numerical Data', in *Auto-Carto 10*, pp. 313–29.

Ramer, U. (1972). 'An Iterative Procedure for the Polygonal Approximation of Plane Curves', *Computer Graphics and Image Processing*, *1*, 244–56.

Raper, J. F., and Bundock, M. S. (1990). 'UGIX: A Layer Based Model for GIS User Interface', in *Proceedings of NATO ASI Cognitive and Linguistic Aspects of Space, Las Navas, Spain* Berlin. Springer-Verlag.

Reisner, P. (1988). 'Query Languages', in *Handbook of Human-Computer Interaction* Amsterdam. North-Holland/Elsevier.

Richardson, D. E. (1988). 'Database Design Considerations for Rule-Based Map Feature Selection', *ITC Journal*, *2*, 165–71.

Richardson, J. E., and Carey, M. J. (1989). 'Persistence in the E Language: Issues and Implementation', *Software: Practice and Experience*, *19(12)*, 1115–50.

Ripple, W. J., and Ulshoefer, V. S. (1987). 'Expert Systems and Spatial Data Models for Efficient Geographic Data Handling', *Photogrammetric Engineering and Remote Sensing*, *53(10)*, 1431–3.

Robinove, C. J. (1986). 'Principles of Logic and the Use of Digital Geographic Information Systems', US Geological Survey.

Robinson, A. H., Sale, R. D., Morrison, J. L., and Muehrcke, P. C. (1984). *Elements of Cartography*, 5th edn., John Wiley, New York.

Robinson, J. T. (1981). 'The K-D-B-Tree: A Search Structure for Large Multidimensional Dynamic Indexes', *ACM SIGMOD*, *10*, 10–18.

Rosenberg, J. B. (1985). 'Geographical Data Structures Compared: A Study of Data Structures Supporting Region Queries', *IEEE Transactions on Computer Aided Design*, *CAD-4(1)*, 53–67.

Roussopoulos, N., and Leifker, D. (1985). 'Direct Spatial Search on Pictorial Databases Using Packed R-Trees', *ACM SIGMOD*, *14(4)*, 17–31.

—— Faloutsos, C., and Sellis, T. (1988). 'An Efficient Pictorial Database System for PSQL', *IEEE Transactions on Software Engineering*, *14(5)*, 639–50.

Rowe, L. A., and Stonebraker, M. R. (1987). 'The Postgres Data Model', Tech. rep. Memorandum no. UCB/ERL M86/85, Computer Science Division, University of California, Berkeley.

Samet, H. (1984). 'The Quadtree and Related Hierarchical Data Structures', *Computing Surveys*, *16(2)*, 187–260.

Samet, H. (1989). *The Design and Analysis of Spatial Data Structures.* Addison-Wesley, Reading, Mass.

Scheifler, R. W. (1989). 'X Window System Protocol, Release 4, X Version 11', Massachusetts Institute of Technology, Cambridge, Mass.

Schilcher, M. (1985). 'Interactive Graphic Data Processing in Cartography', *Computers & Graphics, 9(1),* 57–66.

Seeger, B., and Kriegel, H.-P. (1990). 'The Buddy-Tree: An Efficient and Robust Access Method for Spatial Data Base Systems', in *16th VLDB Conference, Brisbane, Australia* New York. ACM.

Shea, K. S., and McMaster, R. B. (1989). 'Cartographic Generalization in a Digital Environment: When and How to Generalize', in *Auto-Carto 9,* pp. 56–67.

Siemens Data Systems Division (1987). 'SICAD: The Geographical Information System for Modern Mapping', brochure.

Singer, C. (1991). 'SICAD Geographic Database Management: Concepts and Facilities', in *Beiträge zum Internationalen Anwenderforum 1991 Geo-Informationssysteme und Umweltinformatik, Duisburg, 20–21 February 1991* (Berlin, Siemens), pp. 71–83.

Skyum, S. (1990). 'A Simple Algorithm for Computing the Smallest Enclosing Circle', Department of Computer Science, University of Aarhus.

Smith, H. C. (1985). 'Database Design: Composing Fully Normalized Tables from a Rigorous Dependency Diagram', *Communications of the ACM, 28(8),* 826–38.

Smith, T. R., Menon, S., Star, J. L., and Estes, J. E. (1987*b*). 'Requirements and Principles for the Implementation and Construction of Large-Scale Geographic Information Systems', *International Journal of Geographical Information Systems, 1(1),* 13–31.

―――― Peuquet, D., Menon, S., and Agarwal, P. (1987*a*). 'KBGIS-II: A Knowledge-Based Geographical Information System', *International Journal of Geographical Information Systems, 1(2),* 149–72.

Snyder, J. P. (1987). *Map Projections: A Working Manual.* US Geological Survey Professional Paper no. 1395. US Government Printing Office, Washington, DC.

Stansell, T. A. (1983). 'Civil GPS from a Future Perspective', *Proceedings of the IEEE, 71(10),* 1187–92.

Stonebraker, M., and Rowe, L. A. (1986). 'The Design of Postgres', *ACM SIGMOD, 15(2),* 340–55.

―――― ―――― and Hirohama, M. (1990). 'The Implementation of Postgres', *IEEE Transactions on Knowledge and Data Engineering, 2(1),* 125–42.

Straw, A., Mellender, F., and Riegel, S. (1989). 'Object Management in a Persistent Smalltalk System', *Software: Practice and Experience, 19(8),* 719–37.

Stroustrup, B. (1986). *The C++ Programming Language.* Addison-Wesley, Reading, Mass.

Sun Microsystems, Inc. (1987). 'SunINGRES Manual Set.

Sylvester, J. J. (1857). 'A Question in the Geometry of Situation', *Quarterly Journal of Mathematics*, 1, 79.

Tamminen, M. (1983). 'Performance Analysis of Cell Based Geometric File Organizations', *Computer Vision, Graphics, and Image Processing*, 24(2), 160–81.

—— (1984). 'Comment on Quad- and Octtrees', *Communications of the ACM*, 27(3), 248–9.

Tarjan, R. E. (1987). 'Algorithm Design (1986 Turing Award)', *Communications of the ACM*, 30(3), 204–12.

Teunissen, W. J. M. (1988). *HIRASP: A Hierarchical Modelling System for Raster Graphics*. Ph.D. thesis, University of Leiden.

—— and van den Bos, J. (1988). 'HIRASP: A Hierarchical Interactive Rastergraphics System Based on Pattern Graphs and Pattern Expressions', in *Eurographics*, Amsterdam (Berlin, Springer-Verlag), pp. 393–404.

—— —— (1990). *3D Interactive Computer Graphics: The Hierarchical Modelling System HIRASP*. Ellis Horwood series on Computers and their Applications. Ellis Horwood, New York.

—— and van Oosterom, P. J. M. (1988). 'The Creation and Display of Arbitrary Polyhedra in HIRASP', Tech. rep. 88–20, University of Leiden Department of Computer Science.

Thibault, W. C., and Naylor, B. F. (1987). 'Set Operations on Polyhedra Using Binary Space Partitioning Trees', *Computer Graphics*, 21(4), 153–62.

Thoone, M. L. G. (1987). 'CARIN, een navigatie- en informatiessysteem voor auto's', *Philips Technisch Tijdschrift*, 43(11/12), 338–9.

Tomlinson, R. F. (ed.). (1972). *UNESCO/IGU Second Symposium on Geographical Data Handling*, Ottawa, Canada. International Geographical Union Commission on Geographical Data Sensing and Processing.

Tsichritzis, D. C., and Nierstrasz, O. M. (1988). 'Fitting Round Objects into Square Databases', in *ECOOP '88*, pp. 283–99.

Turner, H. (1988). 'Real-Time Data Collection Quality Assurance Models in Three-Dimensional Spatial Information Systems', *International Journal of Geographical Information Systems*, 2(4), 295–306.

Ullman, J. D. (1982). *Principles of Database Systems*. Computer Science Press, Rockville, Md.

Unisys Corporation (1989). 'Open Geographic Information Systems Forum', brochure.

US Bureau of the Census (1970). 'The DIME Geocoding System', Tech. rep. 4, Census Use Study, US Department of Commerce, Bureau of the Census, Washington, DC.

US GeoData (1985). 'Digital Line Graphs from 1:100,000-scale Maps', US Geological Survey, National Mapping Division, Renston, Va.

USA-CERL (1988). 'Geographic Resources Analysis Support System (GRASS) Version 3.0 User Manual', US Army Construction Engineering Research Laboratory, Champaign, Ill.

van de Scheur, M. J., and Stolk, D. J. (1986). 'Informatie- en Rekensysteem ten behoeve van de Rampbestrijding bij Ongevallen met Gevaarlijke Stoffen', Tech. rep. 1986-41, FEL-TNO.

van den Bos, J. (1989a). 'Procol: A Protocol-Constrained Concurrent Object-Oriented Language', *Information Processing Letters, 32*, 221–7.

—— (1989b). 'Procol: A Protocol-Constrained Concurrent Object-Oriented Language. Special Issue on Concurrent Object Languages, Workshop Concurrency, OOPSLA '88, San Diego', *SigPlan Notices, 24(4)*, 149–51.

—— and Laffra, C. (1989). 'Procol: A Parallel Object Language with Protocols', in *OOPSLA '89, New Orleans* (New York, ACM), pp. 95–102.

—— —— (1990). 'Project DIGIS: Building Interactive Applications by Direct Manipulation', *Computer Graphics Forum, 9*, 181–93.

—— —— (1991). 'Procol: A Concurrent Object-Oriented Language with Protocols Delegation and Constraints', *Acta Informatica, 28*, 511–38.

—— van Naelten, M., and Teunissen, W. (1984). 'IDECAP Interactive Pictorial Information System for Demographic and Environmental Planning Applications', *Computer Graphics Forum, 3*, 91–102.

van Oosterom, P. (1988a). 'Spatial Data Structures in Geographic Information Systems', in *NCGA's Mapping and Geographic Information Systems, Orlando, Florida* (Fairfax, Va, National Computer Graphics Association), pp. 104–18.

—— (1988b). 'Spatial Data Structures in Geographic Information Systems', in Hertzberger, L. O. (ed.), *Computing Science in the Netherlands*, pp. 463–77 Amsterdam. SION.

—— (1989). 'A Reactive Data Structure for Geographic Information Systems', in *Auto-Carto 9*, pp. 665–74.

—— (1990a). 'A Modified Binary Space Partitioning Tree for Geographic Information Systems', *International Journal of Geographical Information Systems, 4(2)*, 133–46.

—— (1990b). *Reactive Data Structures for Geographic Information Systems*. Ph.D. thesis, Department of Computer Science, Leiden University.

—— (1991). 'The Reactive-Tree: A Storage Structure for a Seamless, Scaleless Geographic Database', in *Auto-Carto 10*, pp. 393–407.

—— and Claassen, E. (1990). 'Orientation Insensitive Indexing Methods for Geometric Objects', in *4th International Symposium on Spatial Data Handling, Zürich* (Columbus, OH, International Geographical Union), pp. 1016–29.

—— and Laffra, C. (1990). 'Persistent Graphical Objects in Procol', in *TOOLS'90* (Paris, Editions Angkor), pp. 271–83.

—— and van den Bos, J. (1989a). 'An Object-Oriented Approach to the Design of Geographic Information Systems', *Computers & Graphics, 13(4)*, 409–18.

van Oosterom, P. J. M., and van den Bos, J. (1989*b*). 'An Object-Oriented Approach to the Design of Geographic Information Systems', in *Symposium on the Design and Implementation of Large Spatial Databases, Santa Barbara, California* (Berlin, Spinger-Verlag), pp. 255–69.

——— and Vijlbrief, T. (1991). 'Building a GIS on Top of the Open DBMS "Postgres"', in *Proceedings EGIS'91: Second European Conference on Geographical Information Systems* (Utrecht, EGIS Foundation), pp. 775–87.

——— van Hekken, M., and Woestenburg, M. (1989). 'A Geographic Extension to the Relational Data Model', in *Geo'89 Symposium, The Hague* (The Hague, Shape Technical Centre) pp. 319–33.

van Roessel, J. W. (1987). 'Design of a Spatial Data Structure using the Relational Normal Forms', *International Journal of Geographical Information Systems, 1(1)*, 33–50.

——— (1990). 'Attribute Propagation and Line Segment Classification in Plane-Sweep Overlay', in *4th International Symposium on Spatial Data Handling, Zürich* (Columbus, OH, International Geographical Union), pp. 127–40.

van Schagen, P. A. B. (1987). 'Computerwargames voor training en opleiding in de Koninklijke Landmacht', *Militaire Spectator, 156(3)*, 116–22.

Velsink, H. (1990). 'Naar een landelijke bedekking van digitale kaarten voor het kadaster', in *Proceedings MAP '90, De Doelen, Rotterdam* (Hoofddorp, Intergraph), pp. 68–74.

Veltkamp, R. C. (1988). 'The γ-Neighbourhood Graph for Computational Morphology', Department of Computer Science, University of Leiden.

——— (1989). 'A Divide-and-Conquer Algorithm to Compute the 3D Delaunay Triangulation', in *Conference Papers Computing Science in the Netherlands (CSN'89)* (Amsterdam, SION), pp. 463–80.

——— (1990). 'The Flintstones Representation and Approximation Scheme', in *Conference Papers Computing Science in the Netherlands (CSN'90)* (Amsterdam, SION), pp. 485–98.

Vijlbrief, T., and van Oosterom, P. (1992). 'The GEO System: An Extensible GIS', in *Proceedings of the 5th International Symposium on Spatial Data Handling, Charleston, South Carolina* (Columbus, OH, International Geographical Union), pp. 40–50.

Vrana, R. (1989). 'Historical Data as an Explicit Component of Land Information Systems', *International Journal of Geographical Information Systems, 3(1)*, 33–49.

Wagner, M. (1989). 'Structuring Data in Mini-Topo', in *Digital Data Exchange Seminar held at the International Hydrographic Bureau, Monaco, November 1988* (Monaco, International Hydrographic Bureau), pp. 56–70.

Watson, D. F. (1992). *CONTOURING: A Guide to the Analysis and Display of Spatial Data.* Pergamon Press, New York.

Waugh, T. C., and Dowers, S. (1988). 'High-Speed Interactive Mapping in Large Topographic Databases', in *Proceedings of Eurocarto Seven, ITC Publication* (Enschede, ITC), pp. 97–104.

―――― and Healey, R. G. (1987). 'The GEOVIEW Design: A Relational Data Base Approach to Geographical Data Handling', *International Journal of Geographical Information Systems, 1(2)*, 101–18.

Weinand, A., Gamma, E., and Marty, R. (1988). 'ET^{++}: An Object Oriented Application Framework in C^{++}', in *OOPSLA '88* (New York, ACM), pp. 46–57.

―――― ―――― ―――― (1989). 'Design and Implementation of ET^{++}: A Seamless Object-Oriented Application Framework', *Structured Programming, 10(2)*, 63–87.

Werkgroep SUF-2 (1987). 'Standaard – Uitwisselings – Formaat – 2 (SUF-2)', Kadaster, Apeldoorn, Netherlands.

Wiederhold, G. (1986). 'Views, Objects, and Databases', *IEEE Computer, 19(12)*, 37–44.

Wolf, A. (1989). 'The DASDBS GEO-Kernel, Concepts, Experiences, and the Second Step', in *Symposium on the Design and Implementation of Large Spatial Databases, Santa Barbara, California* (Berlin, Springer-Verlag), pp. 67–88.

Wu, P. Y. F., and Franklin, W. R. (1987). 'A Logic Programming Approach to Cartographic Map Overlay', in *Theoretical Foundations of Computer Graphics and CAD*, vol. 40 of *NATO ASI Series F* Berlin. Springer-Verlag.

Wu, J.-K., Chen, T., and Yang, L. (1989). 'QPF: A Versatile Query Language for a Knowledge-Based Geographical Information System', *International Journal of Geographical Information Systems, 3(1)*, 51–7.

Index